Understanding Construction Contracts

Akhtar Surahyo

Understanding Construction Contracts

Canadian and International Conventions

Springer

Akhtar Surahyo
IBI Group
Mississauga, ON, Canada

ISBN 978-3-319-66684-6 ISBN 978-3-319-66685-3 (eBook)
DOI 10.1007/978-3-319-66685-3

Library of Congress Control Number: 2017951455

Printed on acid-free paper

This Springer imprint is published by Springer Nature
The registered company is Springer International Publishing AG
The registered company address is: Gewerbestrasse 11, 6330 Cham, Switzerland

Dedicated to My Mother
Mariam Surahyo
(Deceased, 2005)
Mom, we always remember you and
miss you more than words can express.

Foreword

The construction industry is frequently more risky when compared to other business industries because of the complexity in coordinating various activities. Each project is unique, comprising different means and methods, techniques, and procedures. The core elements of project success are to meet the time, cost, and quality targets. In order to achieve these targets, risk may appear in many ways and could result in time overrun, budget overrun, financial losses, loss of life, environmental damage, and many more failures. This book provides comprehensive information and addresses various topics related to the construction of a contract/project. It also provides awareness regarding how to analyze and manage construction risks to achieve targeted goals.

The book is a practical guide for all those professionals who are involved in the construction process in Canada and internationally and also for undergraduate and graduate engineering students who intend to join the construction industry. Construction professionals from various fields involved in project construction should use this book as an opportunity to understand various clauses quoted from industry standards such as the Canadian Construction Documents Committee (CCDC) and International Federation of Consulting Engineers (FIDIC). The book addresses project cost and time issues, change orders, construction claim issues, construction law, and how to deal with disputes conforming to the requirements of CCDC and FIDIC of construction contract documents.

Contract procurement professionals can use this book to gain a better understanding about the competitive bidding process and information on "Contract A" and "Contract B," qualified bids, and responsive and nonresponsive bids. To reinforce the concepts, examples have been incorporated into the discussions. The book also provides information on what to do about identical low and tied bids and how to apply "best value criteria" between tied bids. This book also provides valuable information on insurance and construction bonds and contract delivery methods and types which will also be useful to contract procurement professionals.

Mr. Surahyo is a practicing professional engineer carrying 30 plus years of hands-on experience with construction contracts and has successfully addressed the topics from the construction perspective, providing thorough knowledge and understanding of every topic within *Understanding Construction Contracts* so as to complete the contract/project successfully meeting the main contract objectives. Mr. Surahyo should be congratulated for providing a practical handbook which will be of great assistance to the construction industry.

I wish Mr. Surahyo good luck and hope this book will go into many future editions.

Graeme King P. Eng., PMP
Senior Project Engineer – Engineering Department
City of Barrie, ON, Canada

Preface

Construction contracts are often very complex with uncertainties, and at any time anything can go wrong. Since every construction contract is unique, each offers a multitude of different issues and problems. Anyone dealing with a construction contract must therefore equip himself with knowledge and understanding of as many areas as practicable regarding construction contracts to complete the contract/project successfully meeting the main contract objectives of scope, time, cost, and quality.

This book encompasses the knowledge and understanding of the author accrued over 30 plus years providing a comprehensive overview of almost all the construction contracting topics in a step-by-step fashion and in an easy and concise form. A distinct feature of this book is that most of the key topics are addressed with particular references to the Canadian standard form of contract document CCDC-2 (Canadian Construction Documents Committee) as well as the international standard form of contract document FIDIC (International Federation of Consulting Engineers) conditions of contract utilizing the construction experience gained over many years by the author.

The book starts by providing basic information on construction contracts and law, explaining contract binding elements with examples including tort law, the Construction Lien Act, and statutory holdback. Chapter 2 describes roles and responsibilities of parties to the contract and their rights and obligations under the standard form of contract CCDC-2 (Canadian Construction Documents Committee) and FIDIC (International Federation of Consulting Engineers) conditions of contract. Chapter 3 describes forms of construction organizations, and Chaps. 4, 5, and 6 provide information on contract delivery methods, types of contracts, and how to select an appropriate contract strategy explaining strengths, weaknesses, and risks of contract methods and types.

Chapter 7 describes Construction Documents and provides information about many important standard forms of Construction Documents, such as documents published by CCDC (Canada), FIDIC (International), JCT (UK), NEC (UK), AIA (USA), and EJCDC (USA). This chapter further provides information on various types of specifications and standard documents "MasterFormat" and "UniFormat" for organizing specifications which are jointly developed by CSC (Construction Specifications Canada) and CSI (Construction Specifications Institute, USA). Chapter 8 highlights bidding procedures describing what is "Contract A" and "Contract B" as well as what is covered under bid documents and how bids are evaluated. This chapter also highlights types of bidding: open bidding, single- and two-stage selective bidding, and negotiated bidding with examples on qualified bids, nonresponsive bids, low or tied bids, unbalanced bids, and best value criteria for selection of a bidder. Chapter 9 describes construction risks and how they are analyzed and managed. Chapter 10 describes construction insurances and bonds such as performance bonds, labor and material payment bonds, and bid bonds. Chapter 11 addresses construction time, briefing commencement and completion dates, extension of time, defects liability period, and delay period and their requirements under CCDC-2 and FIDIC conditions of contract. Chapter 11 further describes how to control the construction time while providing an overview of scheduling methods using MS Project and critical path method complete with examples.

Chapter 12 covers construction costs describing requirements of progress payments, final payments, cash allowances and contingencies, and changes in work under CCDC-2 and FIDIC conditions of contract. Finally, this chapter also describes how to control construction cost. Chapters 13 and 14 describe construction quality and construction safety. Chapter 15 describes construction claims: types of construction claims and how to deal with them in the light of CCDC and FIDIC conditions of contract. This chapter explains in detail about various types of delays: excusable and non-excusable delays and compensable and non-compensable delays including acceleration. Various types of changes are defined for a deeper understanding of the reader: directed changes, constructive changes, concurrent delays, cardinal changes, differing site conditions, and disruption.

Chapters 16 and 17 cover suspension and termination of a construction contract and construction dispute issues with respect to requirements under CCDC-2 and FIDIC conditions of contract. Various modes of contract termination are explained: termination by performance, by breach of contract or default, by fundamental breach, by repudiation breach, by mutual agreement, by unforeseen circumstances (force majeure), and by operation of law. The chapter on construction disputes provides information on dispute resolutions recommended by CCDC and FIDIC conditions of contract. Various dispute resolution modes are discussed including negotiations, mediation, dispute adjudication board, conciliation, arbitration, and litigation. The last chapter describes the Construction Contract Administration and the role of the contract administrator throughout various construction phases.

This book is a practical guide for all those professionals who are involved in the construction process and for undergraduate and postgraduate engineering students who intend to join the construction industry. This book is recommended for inclusion within the Construction Management Syllabus.

Akhtar Surahyo, P.Eng
Mississauga, ON, Canada

Acknowledgments

The author gratefully acknowledges the courtesy extended by Dr. Ing. Francois Baillon of FIDIC (International Federation of Consulting Engineers) and Dianna Fournier of CCDC (Canadian Construction Documents Committee) and others named in the text and references, in permitting the reproduction of the copyrighted material.

The author is most indebted to Peter B. Rozanski, manager of Lakeshore West, from the Capital Projects Group at Metrolinx, Toronto, Ontario, who generously donated his time and provided technical review of the complete book including additional input.

The author would also like to record his gratitude to Graeme King, senior project engineer from the Engineering Department at the City of Barrie, Ontario, for kindly writing the Foreword for this book and for good wishes in his comments.

Additionally, the author would also like to thank Daniel P. Carroll, the partner of Field Law Canada, for the permission to reproduce their complete article in Chap. 1.

The author also greatly acknowledges the following Crown's work:

Source: *Dispute Resolution Reference Guide*, http://www.justice.gc.ca/eng/rp-pr/csj-sjc/dprs-sprd/res/drrg-mrrc/index.html. Department of Justice Canada, July 2006, reproduced in Chap. 17 with the permission of the Department of Justice Canada, 2016.

The author also acknowledges the reproduction of the material in Chap. 14 from the Ontario Occupational Health and Safety Act (OHSA, R.S.O. 1990) and Ontario Regulation 213/91 for Construction Projects (e-Laws Currency Date, January 2016) – Queen's Printer for Ontario, www.ontario.ca/laws. The material referred is an unofficial version of the Government of Ontario legal materials.

The author also acknowledges the reproduction of the material in Chaps. 1, 10, 11, and 12 from the Ontario Construction Lien Act (R.S.O. 1990) last amended on 2010 – Queen's Printer for Ontario, www.ontario.ca/laws. The material referred is an unofficial version of the Government of Ontario legal materials.

The author also acknowledges the reproduction of the material in Chaps. 12, 13, and 15 from the Ontario Provincial Standards for Roads and Public Works – Queen's Printer for Ontario 2008 (last amended on January 2016). The material referred is an unofficial version of the Government of Ontario legal materials.

Contents

1 Construction Contracts and Law Basics 1
- 1.1 Introduction 1
- 1.2 Types of Law 1
- 1.3 Contract Definition 2
- 1.4 Contract Elements 2
  - 1.4.1 The Offer or Proposal 3
  - 1.4.2 The Acceptance 3
  - 1.4.3 Mutual Intent 4
  - 1.4.4 Consideration 4
  - 1.4.5 Capacity 4
  - 1.4.6 Legality 4
- 1.5 Tort Law 5
  - 1.5.1 Categories of Torts 6
- 1.6 The Construction Lien Act 7
- 1.7 Statutory Holdbacks 8
  - 1.7.1 Basic Holdback 8
  - 1.7.2 Finishing Holdback 8
  - 1.7.3 Placing and Enforcing a Construction Lien in Ontario 9
- 1.8 Environmental Law 10
- 1.9 Quantum Meruit 11
- 1.10 Express Terms 11
- 1.11 Implied Terms 12
- 1.12 Bid Shopping 12
- 1.13 Exclusion Clause 13
- 1.14 Privilege Clause 15
- 1.15 Rescission of the Contract 15
- 1.16 Contra Proferentem 16
- References and Further Reading 16

2 **Parties to the Contract** 17
2.1 Introduction 17
2.2 The Owner 18
2.2.1 Rights and Obligations of Owner 19
2.3 The Consultant 21
2.3.1 Rights and Obligations of the Consultant 22
2.4 The Contractor 23
2.4.1 Rights and Obligations of the Contractor 24
Reference and Further Reading 27

3 **Forms of Construction Organizations** 29
3.1 Introduction 29
3.2 The Sole Proprietorship 29
3.3 The Partnership (General) 30
3.3.1 Limited Partnership 31
3.4 Corporations 31
3.5 The Joint Venture (JV) 33
3.5.1 Basic Elements of a Joint Venture 33
3.5.2 Termination of Joint Ventures 34
References and Further Reading 34

4 **Contract Delivery Methods** 35
4.1 Introduction 35
4.2 Contract Delivery Methods 36
4.2.1 Design-Bid-Build Method 36
4.2.2 The Design and Build Method (DB) 37
4.2.3 Other Forms of Design Build Concept 38
4.2.4 The Construction Management Method 40
4.2.5 Single and Multiple Prime Contract Technique 42
References and Further Reading 44

5 **Types of Construction Contracts** 45
5.1 Introduction 45
5.2 Price-Based Contracts 45
5.2.1 Lump Sum Contract 45
5.2.2 Unit Price Contract 47
5.3 Cost Reimbursable Contracts 48
5.3.1 Cost Plus Percentage Fee Contract 48
5.3.2 Cost Plus Fixed Fee Contract 48
5.3.3 Cost Plus Variable Fee Contract 49
5.3.4 Guaranteed Maximum Cost Contract (GMC Contract) 49
5.3.5 Time-and-Material (T&M) Contract 51
5.3.6 Strengths and Weaknesses of Cost plus Contracts 51
References and Further Reading 52

6 Selection of an Appropriate Contract Strategy 53
6.1 Introduction 53
6.2 Project Scope 53
6.3 Cost 54
6.4 Time 55
6.5 Quality 55
6.6 Size of Project 55
6.7 Owner Requirements 56
6.8 Risk Allocation 56
6.9 Owner Expertise and Responsibilities 59
References and Further Reading 59

7 Construction Documents 61
7.1 Introduction 61
7.2 Contract Forms 62
7.2.1 Form of Agreement 62
7.2.2 Performance Bond 62
7.2.3 Labor and Material Payment Bond 63
7.2.4 *WSIB* Declaration 63
7.2.5 Insurance Certificate 64
7.3 Conditions of Contract 64
7.3.1 General Conditions 64
7.3.2 Particular/Supplementary Conditions 64
7.4 Standard Forms of Conditions of Contract 65
7.4.1 Canadian Construction Documents Committee (CCDC) 65
7.4.2 FIDIC: International Federation of Consulting Engineers 67
7.4.3 NEC (New Engineering Contract) Contracts, UK 68
7.4.4 Infrastructure Conditions of Contract, UK 69
7.4.5 JCT (Joints Contracts Tribunal) Contracts, UK 69
7.4.6 Engineers Joint Contract Documents Committee (EJCDC) Documents, USA 70
7.4.7 American Institute of Architect (AIA) Documents, USA 70
7.5 Specifications 71
7.5.1 Performance Based 72
7.5.2 Prescriptive Based 72
7.6 Drawings 74
7.7 Bill of Quantities (B of Q) 75
7.8 Addendum 76
References and Further Reading 77

8 Bidding Procedures . . . 79
8.1 Introduction . . . 79
8.2 Contract "A" and Contract "B" . . . 80
8.3 Bid Documents . . . 80
8.4 Bidding Requirements . . . 81
8.4.1 Invitation to Bid/Bid Solicitation . . . 81
8.4.2 Instructions to Bidders . . . 82
8.4.3 Information Available to Bidders . . . 82
8.4.4 Bid Security or Bid Bond . . . 85
8.4.5 Bid Forms . . . 85
8.4.6 Supplements to Bid Forms . . . 86
8.5 Bidding Procedures . . . 86
8.5.1 Competitive Bidding . . . 86
8.5.2 Negotiated Bidding . . . 88
8.5.3 Electronic Bidding . . . 89
8.6 Receiving and Opening of Bids . . . 89
8.7 Bid Evaluation . . . 90
8.7.1 Administrative Evaluation . . . 90
8.7.2 Technical Evaluation . . . 92
8.7.3 Financial Evaluation . . . 95
8.8 Award of Contract . . . 95
References and Further Reading . . . 96

9 Construction Risk Analysis and Management . . . 97
9.1 Introduction . . . 97
9.2 Definition of the Risk . . . 97
9.3 Risk Management . . . 98
9.3.1 Risk Management Planning . . . 98
9.3.2 Identifying the Risk . . . 99
9.3.3 Analyzing the Risk . . . 99
9.3.4 Responding and Allocating Risk . . . 102
9.3.5 Risk Monitoring and Control . . . 105
9.4 Standard Forms of Contract . . . 105
References and Further Reading . . . 106

10 Construction Insurance and Construction Bonds . . . 107
10.1 Construction Insurance . . . 107
10.1.1 Introduction . . . 107
10.1.2 Insurance Policy . . . 108
10.1.3 Basic Concepts . . . 108
10.1.4 Contract Requirements . . . 111
10.1.5 Types of Insurance Coverage . . . 114
10.2 Construction Bonds . . . 119
10.2.1 Introduction . . . 119
10.2.2 Types of Bonds . . . 120
References and Further Reading . . . 124

11 Construction Time 127
11.1 Introduction 127
11.2 Commencement Date. 127
11.3 Completion Date 128
11.3.1 Substantial Performance of the Work 129
11.3.2 Total Performance 129
11.4 Extension of Time 130
11.5 Liquidated Damages 131
11.6 Controlling Construction Time 132
11.6.1 Scheduling Methods. 133
11.6.2 Bar Chart or Gantt Bar Chart. 134
11.6.3 Critical Path Method (CPM) 137
References and Further Reading. 142

12 Construction Payment/Cost. 143
12.1 Introduction 143
12.2 Contract Price. 143
12.3 Progress Payments 144
12.4 Final Payments. 145
12.4.1 Procedures Under FIDIC Conditions of Contract 145
12.4.2 Procedures Under CCDC-2 and Related Canadian Documents 146
12.5 Cash Allowances and Contingencies 148
12.5.1 Cash Allowances 148
12.5.2 Contingency Allowance 148
12.6 Changes in the Work 149
12.6.1 Dealing with Changes 150
12.6.2 Valuation of Changes/Variations 150
12.7 Construction Cost Control 151
12.8 Forecasting. 155
12.9 Value Engineering 156
12.10 Life Cycle Costing (LCC) 157
Reference and Further Reading 159

13 Construction Quality 161
13.1 Introduction 161
13.2 Contract Requirements. 162
13.3 Quality Management 162
13.4 Quality Management Processes 163
13.4.1 Quality Planning 163
13.4.2 Quality Control (QC) 164
13.4.3 Quality Assurance (QA). 165
13.5 QA, QC Implementation 166
13.5.1 Quality in Materials 166
13.5.2 Quality in Workmanship 167
References and Further Reading. 177

14 Construction Safety ... 179
14.1 Introduction ... 179
14.2 Contract Requirements ... 179
14.3 Common Safety and Health Issues and Guidelines ... 181
14.3.1 Personal Protective Equipment (PPE) ... 182
14.3.2 Protection from Falls and Falling Objects ... 182
14.3.3 Protection from Machinery Hazard ... 183
14.3.4 Protection from Trenching Hazard ... 184
14.3.5 Protection from Confined Space Hazards ... 186
14.3.6 Protection from Electrical Hazard ... 187
References and Further Reading ... 188

15 Construction Claims ... 189
15.1 Introduction ... 189
15.2 Construction Claims ... 189
15.2.1 Definition of a Claim ... 189
15.3 Types/Causes of Claims ... 190
15.4 Dealing with Claims ... 194
15.5 Maintaining Claim Records ... 196
15.5.1 Types of Records Required ... 196
15.6 Pricing of Claims ... 198
15.6.1 Direct and Indirect Costs ... 198
15.7 Claim Avoidance ... 201
Reference and Further Reading ... 204

16 Suspension and Termination of Construction ... 205
16.1 Introduction ... 205
16.2 Contract Termination and Discharge ... 205
16.3 Termination by the Owner Under FIDIC Conditions of Contract ... 207
16.3.1 Costs and Risks of Terminating the Contract ... 209
16.3.2 Termination by the Employer for Convenience (Without Cause) ... 210
16.4 Termination by the Owner Under CCDC-2 Conditions of Contract ... 210
16.5 Termination by the Contractor Under FIDIC Conditions of Contract ... 211
16.6 Suspension of Construction Under FIDIC Condition of Contract ... 212
16.7 Termination and Suspension by the Contractor Under CCDC-2 ... 213
References and Further Reading ... 214

17 Construction Disputes 215
17.1 Introduction 215
17.2 CCDC-2 Dispute Resolution Procedure 215
17.2.1 Negotiation 216
17.2.2 Mediation 216
17.2.3 Arbitration 216
17.3 FIDIC Dispute Resolution Procedure 217
17.3.1 The Dispute Adjudication Board (DAB) 217
17.3.2 Amicable Settlement 220
17.4 Arbitration 223
17.5 Litigation 224
References and Further Reading 224

18 Construction Contract Administration 225
18.1 Introduction 225
18.2 Contract Administrator 226
18.3 Project Phases 227
18.3.1 Preconstruction Phase 227
18.3.2 Construction Phase 227
18.3.3 Post-construction Phase and Project Closeout 234
18.4 Project Warranties 237
18.4.1 Express Warranties 238
18.4.2 Implied Warranties 239
18.4.3 Extended Warranties 239
Reference and Further Reading 241

Index 243

About the Author

Akhtar Surahyo is a graduate civil engineer with a postgraduate diploma in structural engineering. He is a registered professional engineer with Professional Engineers Ontario, Canada, and a member of the Project Management Institute, USA. He has extensive practical construction experience of more than 30 years in executing and managing multidisciplinary projects in many countries and has successfully completed many projects from inception through completion. Mr. Surahyo is also author of the book *Concrete Construction: Practical Problems and Solutions* published by Oxford University Press in 2002. Currently, Mr. Surahyo is working as a contract administrator with IBI Group Canada.

Chapter 1
Construction Contracts and Law Basics

1.1 Introduction

A construction contract is a legally binding agreement, usually between two parties – the Owner and the Contractor. Construction contracts are usually made in writing. The Owner is the party who makes a decision to carry out a project with his requirements and pays the Contractor for the project. In return, the Contractor is obligated to construct the project in accordance with the Contract Documents. In most cases, the Owner hires Consultants who prepare the Construction Documents and also administer the contract in the capacity of an agent of the Owner.

The Contract Documents include the details of work, which the Contractor has offered to construct and the payments which the Owner will pay to the Contractor. The Contract Documents also include "conditions of contract" which specify the responsibilities and obligations of both parties. When unexpected and unforeseen problems arise, the "conditions of contract" determine which of the parties must bear the consequences of delays and additional costs. Among other items Construction Documents also include specifications, drawings, bill of quantities, etc.

The Contractor may be selected on the basis of any type of competitive bidding, or the Owner may negotiate a contract with a selected Contractor. The entire project may be included within a single general contract or separate prime contracts for specific portions of the job. All aspects of construction contracts are discussed in the following sections.

1.2 Types of Law

In order to become fully aware of the rights and obligations under the contract, the construction professionals must understand the basic principles of Contract Law.

A. Surahyo, *Understanding Construction Contracts*,
DOI 10.1007/978-3-319-66685-3_1

Briefly, law may be classified as either Public or Private. Public Law mainly deals with the rights and obligation of government, whereas, Private Law deals with rights and obligations of individuals or private organizations. Public and Private Law mainly cover [1]:

1. *Public Law*
    - Constitutional law
    - Administrative law
    - Criminal law
2. *Private Law*
    - Contract law
    - Tort law
    - Property law

Sometimes decisions are made on a common law basis by courts. Common law means common to whole populations and refers to the legal principles established in previous court decisions (the theory of precedent) or judge-made law.

1.3 Contract Definition

Standard Construction Document CCDC-2 (Canadian Construction Document Committee) defines a “contract” as an undertaking or agreement by two or more parties to perform their respective duties, responsibilities, and obligations as prescribed in the Contract Documents and represents the entire agreement between the parties [2]. Contract Documents attached to or stated in the agreement form integral parts of the contract. A construction contract consists of mutual promises exchanged by the parties to that contract. A contract is a binding legal agreement; once made, it cannot be easily set aside or varied.

1.4 Contract Elements

To make a contract enforceable and binding, the presence of following five elements is important [1]:

1. Offer and acceptance
2. Mutual intent
3. Consideration
4. Capacity to contract
5. Legality

1.4.1 The Offer or Proposal

An offer is basically a promise made by one party who agrees to perform a certain task in accordance with specific terms and conditions. The offer may be made verbally; however, it is always preferable to document the agreement in writing. For this reason, construction contracts are mostly made in writing, using one of the standard formats.

Primarily, an offer cannot be withdrawn or revoked by the party who made it for a specific time. For example, in construction, bidders submit tenders and make it irrevocable for a specific period of time upon instructions from the Owner. If the offer is not accepted within a reasonable time, it will lapse, i.e., it will become void. In construction contracts, most of the consultants/owners demand certain procedures to be followed while submitting an offer, like specific time, validity of date, sealed envelopes, etc. On noncompliance such as incomplete bids, qualified bids, and/or nonresponsive bids, the offer is rejected. Bid noncompliance is further described in Chap. 8.

Sometimes, parties may agree to negotiate on the basis of the offer. For example, the Owner may intend to add something to the tender or may intend that the work should be completed 2 months earlier than decided; such issues may result in revision of the tender. A fresh offer submitted based on additional requirements is known as a counter offer and will be subjected to the revised conditions applied.

At any time prior to acceptance, the offer may be revoked through a notice unless the offering party is bound by an irrevocable clause. An offer can also expire if the deadline for acceptance is over.

1.4.2 The Acceptance

Once an offer has been provided, it will become a binding contract; if it is accepted, it means that it has been executed or signed by the other party. The acceptance of offer must be made clear/unambiguous and unconditional; otherwise it will not be treated as a binding offer or acceptance.

As an example [Peter Lind & Co. Ltd. v. Mersey Docks & Harbour Board (1972)], an offer to build a freight terminal was made by a tender. The Owner purported to accept "your tender" on receipt of two alternative bids for a project, one on a fixed price and the other was on a varying price dependent on labor and materials' costs (time and material). Since the Owner did not clarify which contract, the court held that there was no contract due to lack of certainty [6].

1.4.3 Mutual Intent

Mutual intent means that the parties involved must have a meeting of minds and decide that they agree to do the things as mentioned in the contract. For this reason, usually a letter of intent is issued that expresses interest in proceeding with negotiation toward a contract. Letter of intent is the informal acceptance of the contract and does not make a binding until the formal "letter of acceptance" is issued.

1.4.4 Consideration

Each party who negotiates a contract promises something in return for the other party's undertakings. Consideration can be described as something of value that is exchanged by contract parties [4]. The consideration given by each party does not have to be of equal value. It is the basic requirement of the contract that the agreement must be supported by consideration. Consideration may be in the shape of payment of money, performance of work, supplying some goods, etc. A promise made by a Contractor to construct a house in return for a payment of a fee is an example of consideration. Each party must get something in return for his promise; otherwise there is no consideration. No contract is valid where consideration is not present. Additionally, the consideration should not conflict with established law and should not relate to some event in the past.

1.4.5 Capacity

A contract should only be made by parties capable of fulfilling their intended role. Under the law, not everyone has the necessary capacity to enter into an agreement. A party to an agreement can be considered to lack the capacity resultant from lunacy, drunkenness, mental disorder, etc. Similarly minors are not bound by their contracts; however, they are responsible for the value of goods received in contracts made for necessities of life like food, shelter, clothing, etc.

Corporations also cannot enter into a binding contract if it is beyond their stated powers. Hence, when dealing with corporations, the Engineer must be careful to determine the nature of the corporation. The contract will not be legally enforceable if it is beyond the stated powers of the corporation.

1.4.6 Legality

If the purpose of the contract is unlawful, i.e., illegal, it will not be enforced. For example, a contract violating municipal regulations is not binding and is void in courts. A void contract creates no legal rights and therefore cannot be sued upon. A contract must also be made with free consent of all parties. Consent is said to be free when it is not caused by force or undue influence, fraud, or misrepresentation. Thus, if a contract is procured by duress or fraud, it will be treated as void.

A contract will also be said unenforceable if it is one-sided, where the terms are unreasonably favorable to one party. Unenforceable contracts are in fact valid; however, due to omission of required formalities, a party seeking to enforce it will be denied a remedy [3].

To make a contract enforceable, it should also be free from immoral or criminal purpose and should not contain such clauses that are against public policy.

1.5 Tort Law

In French, Tort means wrong. In legal terms, Tort refers to a wrongful act where one person causes damage, injury, or harm to another person or property for which the law provides a remedy. The wrongful act may result from intentional actions or a breach of duty as in negligence or due to a violation of statutes. Tort does not fall under criminal law; it is a civil wrong and is different from a breach of contract.

In fact, when the Engineer is asked for his advice and when he knows that his skill and judgment were relied upon and he was trusted to give advice, he became bound to take a reasonable degree of care, skill, and diligence in exercising his undertaking as an advisor. He becomes liable for any damages that result from any advice he gave because of incompetence, carelessness, or negligence. The concept of Tort law is to provide relief for damages incurred and deter others from committing the same harms.

For Tort liability to arise, no privity of contract is necessary. (Privity of contract describes the legal relationship between those who are direct parties to it; it is further explained in the following pages.) For example, no contract exists between a negligent driver and the victim of the car accident. Even if there is no contract between the parties, liability in contract will arise if the services are performed negligently by professionals. Therefore, most Engineers are required to get professional liability insurance that provide protection if an Engineer's negligence causes harm resulting in Tort.

In order for Tort liability to be warranted and to satisfy the court for compensation, the plaintiff would need to show the following fundamentals of negligence:

1. A duty of care must exist between the defendant and the plaintiff.
2. Conduct of the defendant breached that duty of care.
3. The defendant's breach caused the damage to the plaintiff.

To satisfy a court, all conditions above must be met. Among the types of damages the injured party may recover are: loss of earning capacity, pain and suffering, and reasonable medical expenses. They include both present and future expected losses.

1.5.1 Categories of Torts

There are numerous specific forms of Tort liability including trespass, assault, battery, negligence, product liability, and intentional infliction of emotional distress.

However, Tort law can be classified into three general categories: intentional Tort, negligent Tort, and strict liability Tort. There are also separate areas of Tort law including nuisance, defamation, invasion of privacy, and a category of economic Tort:

- *Intentional Tort* – Intentional Tort refers to harms done to people intentionally by the willful misconduct of another, such as assault, fraud, and theft. Hitting another person in a fight is an intentional act that would be the Tort of battery. Striking a person accidentally would not be an intentional Tort since there was no intent to strike the person.
- *Negligence Tort* – Negligence can exist due to the failure to exercise a reasonable standard of care for another person. Causing an accident by failing to obey traffic rules and professional negligence are good examples of negligent Torts. Negligence constitutes the majority of Tort claims [11].

Duty of care and standard of care are of primary concern to risk and liability. Professionals must exercise a standard of care that is considered as "average." Most negligence actions turn on the issue of whether the defendant had breached the standard of care owed to the plaintiff [8]:

- *Strict liability Tort* – Strict liability Tort or nuisance unlike negligence and intentional Tort is not concerned with the person's harmful act. Instead, strict liability focuses on the act itself: if someone commits a certain act (e.g., making and selling defective products), then that person is responsible for the damages from that act regardless of the level of care exercised or their intentions.

Another example could be a Contractor during construction using dynamite to blast some rocky strata, which causes debris to be thrown onto the land of another person and damages his house. The house Owner may recover damages from the Contractor even if the Contractor was not negligent and did not intend to cause any harm. This is called strict liability or absolute liability:

- *Vicarious liability* – Vicarious liability refers to a situation where someone is held responsible for the actions or omissions of another person. While performing duties during the course of employment, if an Employee commits an act of Tort, the Employer will be vicariously liable for the damage caused. Vicarious liability is sometimes referred to as strict, or no-fault, liability because the Employer himself is not personally at fault.

Under vicarious liability, the Employer provides compensation to the injured party because the Employer is considered to be in better financial position than the Employee. However, in some cases along with the vicarious liability of the Employer, the Employees are also found liable for the Tort committed by them.

Hence, to protect its Employee Engineers, a company providing engineering services usually obtains professional liability insurance for both the company and its Employee Engineers.

1.6 The Construction Lien Act

The construction lien is a legal right created by statute and, in Ontario, Canada, is governed by the Construction Lien Act. Section 14(1) of Chapter C30 of the act specifies that [7]:

> *A person, who supplies services or materials to an improvement for an Owner, Contractor, or Subcontractor, has a Lien upon the interest of the Owner in the premises improved for the price of those services or materials.*

In the construction industry, most construction work is carried out by different construction trades who have no privity of contract with the Owner of the project and who work without any form of security. Ontario's Construction Lien Act represents an important exception to the privity of contract rule.

Privity of contract describes the legal relationship between parties to a contract. The rule of privity is that the rights and obligations created by a contract are enforceable by and against the parties to that contract only. A contract can neither impose obligations nor confer rights upon others who are not party to it. Privity of contract law prevents a sub-trade who provides services or supplied materials used in the project from asserting a claim against the Owner of the premises, unless that person also had a contract directly with the Owner. The purpose of the Lien Act is to avail the lien to all persons who supply services or materials to a particular construction project or improvement, even if there is not any direct contractual agreement between such persons and the Owner. The Construction Lien Act sets out the rules as to who has a lien (security against the property) and the process by which lien claimants can enforce their various rights.

For the purpose of the act, the terms "improvement," "Owner," "Contractor," "Subcontractor," and "premises" are defined as having particular and unique meanings. For example, the "improvement" in respect of any land is defined in Section1(1) of the Ontario's Lien Act as [7]:

(a) Any alteration, addition, or repair to the land
(b) Any construction, erection, or installation on the land, including the installation of industrial, mechanical, electrical, or other equipment on the land or on any building, or structure or works on the land that is essential to the normal or intended use of the land, building, structure, or works,
(c) The complete or partial demolition or removal of any building, structure, or works on the land

Persons entitled to a lien normally include contractors, subcontractors, workers, and material and equipment suppliers. Engineers and architects who prepare designs for an improvement which is never undertaken are also entitled to a lien. An Ontario case determined that legal services do not come within the definition of "supply of services" under the Ontario's Lien Act [10].

A person's lien comes in effect when he first supplies services or materials to the construction project. A lien can expire unless it is preserved. All real property under improvement is subject to a lien except the land of the federal or provincial crown, public street, and highway or railway right of way. The lien held by a party is limited to the unpaid amount of the services or materials supplied to the project and is protected by the holdback requirements of the act and contract.

1.7 Statutory Holdbacks

The Ontario's Lien Act prescribes two primary types of holdbacks: basic and finishing.

1.7.1 Basic Holdback

Part IV of The Ontario's Lien Act [7] provides that each payer (Owner, Contractor, or Subcontractor) upon entering into a contract or subcontract is required to retain a holdback equal to 10% of the price of the services or materials as they are actually supplied under the contract or subcontract until all liens that may be claimed against the holdback have expired or have been satisfied or discharged or vacated by payment into court.

1.7.2 Finishing Holdback

In addition to basic holdback, the act also calls for a separate holdback for finishing works. The finishing holdback will also be 10% of the price of the remaining services or materials after substantial performance until all liens that may be claimed against the holdback have been dealt with as mentioned above.

On receipt of any written notice of a lien claim, the payer must retain an amount sufficient to satisfy the full amount of the lien under notice in addition to 10% holdback.

It may be noted that the holdback is not intended to be retained by the Owner for deficiencies or incomplete work [12]. The purpose of the holdback is to cover any liens that are filed against the project.

1.7.2.1 Time Limits

Holdback monies can only be released when all lien rights have expired. All construction liens expire 45 days after the contract is complete unless a claim for lien is preserved by registering a claim against title. Therefore, the Owner should have their lawyer review the title of the property on the 46th day after the contract is complete. If no claims for lien have been registered against the title on or before the 46th day after the contract is complete, then the Owner should pay the basic holdback to the Contractor. If a claim for lien is made before the 46th day, then the Owner should not pay anything further to the Contractor and seek further advice from their lawyer.

The Ontario's Lien Act provides a set of time limits for the expiration of lien rights. With respect to a basic holdback of a General Contractor, lien rights expire after 45 days following the earliest of one of the following events:

1. The date the contract is completed.
2. The date the contract is abandoned.
3. The date on which a copy of the certificate of substantial performance is published in a construction newspaper.

Similarly a Subcontractor has also 45 days to register a claim for lien against the property from the earliest of the last date of supply of services or materials to the project or certification of completion. After this 45-day period, all lien rights of the Subcontractor expire, and the holdback funds can be released.

Finishing holdback obligations comes to an end under the same conditions except that the expiry of the lien period is 45 days after total completion of the project. If at any point the project is abandoned, the expiry of the lien period is 45 days from the date that the project was abandoned.

1.7.3 Placing and Enforcing a Construction Lien in Ontario

Considering an example, where a Subcontractor has supplied work or material to a Contractor and has not been paid, then that Subcontractor has a right to register a claim for lien against the Owner's interest in the land.

In Ontario, a lien holder must take action to preserve their lien within 45 days (as mentioned above) of the date the contract is completed or abandoned. The lien is preserved by registering a claim for lien on the title of the improved land. This type of lien is called "lien attached to premises." The other type of lien is "lien not attached to premises." In this case, the lien is preserved by giving the Owner a copy of the claim for lien.

The claim for lien is a standard-form document in Ontario. The claim for lien will need to include a description of the services or materials supplied, the amount claimed, and a description of the premises to which the lien relates. Once the lien is preserved, then it is required to perfect the lien within 45 days, meaning to file a statement of claim with the court requesting a remedy.

The Owner, on the other hand, does have certain obligations to protect the subcontractors. Those obligations are found in the holdback provisions of the Construction Lien Act, where some funds must be retained and not be paid to the contractor for a period of time. The basic holdback (as mentioned above) is 10% of the value of the work supplied. However, if the subcontractor gives proper notice to the Owner, then the holdback might be increased by the amount actually owed by the owner to the contractor, till the court pronounces the judgment.

1.8 Environmental Law

Companies doing business in Canada should identify and manage the risks associated with legal liability for harm that has been (or may be) caused to the environment. Environmental legislation (statutes and regulations) in Canada are made and enforced by all three levels of government: federal, provincial/territorial, and municipal. Prior to starting a project in Canada, it typically requires approvals (permits, licenses, or registrations) for activities that can impact or relate to the environment, such as discharges into the atmosphere or water. Failure to obtain or comply with a necessary approval is an offence. Such offences can result in significant fines and penalties against a company, as well as individuals, including a company's officers and directors.

Most environmental statutes also authorize environmental agencies to issue administrative orders to a broad range of parties. Further, there are civil actions that can be commenced, based on legislation or common law, for matters such as damage caused by contamination. These actions are typically based on breaches of legislation or the common law actions of nuisance, negligence, trespass, and strict liability [13].

Certain government initiatives, as well as designated private projects, may be subject to a requirement to conduct a provincial and/or federal environmental assessment (essentially an assessment of the potential effects of the proposal or activity on the environment). An environmental assessment may need to be completed prior to the issuance of project-related licenses or government funding or prior to the disposition of crown lands. It is important for project proponents to fully assess and mitigate scheduling, regulatory, and litigation risks associated with environmental assessments.

Some of the other common law terms used in construction are briefed hereunder for understanding.

1.9 Quantum Meruit

In the law of contracts, quantum meruit is a doctrine by which the law infers a promise to pay a reasonable amount for labor and materials furnished, even in the absence of a specific legally enforceable agreement between the parties. Quantum meruit is basically a reasonable payment for work that has been earned. Sometime, an Owner requests a Contractor to perform certain services, and no payment terms are finalized in return. In such cases, a court can award payment on a quantum meruit basis, i.e., on the basis of reasonable amount of compensation for the work completed.

A claim on a quantum meruit basis is appropriate where there is an implied agreement by the Owner to pay a reasonable sum in return for the work completed. For example, most professional roofers hired to repair a roof insist on having a formal agreement with the Owner of the house before beginning the repairs. In the absence of an agreement or formal contract, the roofer may be unable to recover losses in court if the transaction goes awry. Quantum meruit is a judicial doctrine that allows a party to recover losses in the absence of an agreement or binding contract.

Quantum meruit can be used to address situations where no contract exists or where a contract exists but for some reason is unenforceable. In such cases, courts imply a contract to avoid an unjust result. Such contracts are called quasi contracts. In assessing a quantum meruit claim, various suitable methods of valuation can be adopted. For example, value may be based upon the costs of labor and materials plus profit or the measurement of the work based on reasonable rates.

1.10 Express Terms

The terms written into the Contract Documents or oral promises leading to the signing of a formal contract are treated as "express terms." It is always preferable to write a contract specifying all agreed terms by the parties. In cases of dispute with oral agreements, which do not comprise part of the written contract, a degree of uncertainty may exist because it may be treated as a term of the contract depending on whether it was made fraudulently, negligently, or innocently.

Primarily, the outlined terms used in the contract form are Conditions of Contract and warranties. Conditions are considered the major terms of the contract, and, for its breach, the remedies of repudiation or rescission of the contract and damages are permitted. Whereas, warranties are considered the minor terms of the contract, and, for its breach, only damages are permitted.

1.11 Implied Terms

Those terms, which the parties overlooked and did not expressly stated in writing or conduct to include making a contract, are termed as "implied terms." These terms do not appear in the parties' agreement but are incorporated into construction contracts by operation of law when required by the substantive law governing the parties' agreement. Such terms with respect to parties' duties and obligations normally include the obligation of both parties to deal with each other fairly and in good faith. For example, the Owner's duty is to cooperate with the Contractor in completing the work, and the duty to disclose information material to the Contractor's performance and not to hinder construction work, and the duty to provide accurate plans and specifications. They also include the Contractor's implied obligation to comply with construction industry standards and customs and duty to complete the project in a good and workmanlike manner.

Implied terms in a contract are terms that form part of the contract even though they are not expressly included in the body of the contract. The breach of these implied duties and obligations can provide the Contractor with a remedy in the form of additional compensation and a defense to Owner claims for construction delays or deficiencies in the work. Similarly, a breach of the implied duties and obligations of the Contractor can provide the Owner with a defense or an independent claim for damages.

1.12 Bid Shopping

Bid shopping is described as a practice of disclosing a Contractor's or Subcontractor's bid to other potential contractors or subcontractors before the award of a contract in order to drive prices down so as to get a lower bid. It is often regarded as an underhanded practice and can be considered as a breach of Contract "A" law of competitive bidding. To minimize bid shopping practice, one of the requirements of the competitive bidding process is that the main or general contractors should submit their list of subcontractors with their bids to owners. The General Contractor cannot change subcontractors if its bid is accepted, unless a Subcontractor is unable or unwilling to perform and the General Contractor must seek permission from the bid-calling authority for substitution.

In one case Naylor Group Inc. *v*. Ellis-Don Construction Ltd., [2001] 2 S.C.R. 943, 2001 SCC 58, the Oakville-Trafalgar Memorial Hospital ("OTMH") called for tenders for the construction of an addition and renovation of its hospital through the Toronto Bid Depository. Ellis-Don Construction Ltd., one of the largest construction firms in Ontario, approached Naylor Group Inc. in November 1991 to bid for the electrical work on the project.

Naylor's bid had been carried by Ellis-Don in its bid to the Owner. Ellis-Don was ultimately selected by the Owner as the Prime Contractor for the project. Before submitting its bid through the bid depository, Naylor had advised Ellis-Don that its workers were not affiliated with the International Brotherhood of Electrical Workers ("IBEW"). Ellis-Don was, at the time, involved in an Ontario Labour Relations Board ("OLRB") dispute involving the IBEW but informed Naylor that this would not be a problem. However, after the submission of tenders and before the prime contract was awarded to Ellis-Don, the OLRB ruled that Ellis-Don was required to use only electrical subcontractors whose employees were affiliated with the IBEW. Ellis-Don, in turn, advised Naylor of this and offered to subcontract the electrical work to Naylor only if it would align itself with the IBEW, which Naylor declined. Ellis-Don then awarded the electrical work to another Subcontractor, which resulted in Naylor's action for breach of contract [9].

Naylor sued for breach of contract and unjust enrichment. Its contractual claim was dismissed but the trial judge, out of an abundance of caution, assessed the damages that would have been awarded per the contract, if the claim had succeeded, at $730,286. He did allow a claim for unjust enrichment in the amount of $14,560, which corresponded to the costs of preparing the bid, whereas, the Supreme Court of Canada subsequently concluded that these actions constitute a bid shopping and Naylor was awarded damages for breach of contract in the amount of $182,500, rather than the initial trial judge's award of the relatively small costs to Naylor of preparing the bid.

1.13 Exclusion Clause

Exclusion clauses are types of exemption clauses used in contracts. Generally, such clauses attempt to exclude Tort liability or liability for breaches of the contract in which they are contained. An exclusion clause prohibits or excuses one party's liability completely for specified outcomes. These clauses usually provide that one party to the contract will not be responsible for certain outcomes should they occur. For example, if you park your car in a public car park for a fee, the Owner generally includes a provision in the contract that they will not be responsible for damage to your vehicle or theft of goods from it while it is in the car park. A party relying on an exclusion clause is required to draft it properly using clear and unambiguous language so that it exempts the parties from liability arising out of it. In case of ambiguity in terms, the courts usually interpret the provision strictly against the party relying on the clause based on "contra proferentem" principle explained in the following pages.

Tercon Contractors Ltd. *v.* British Columbia (Transportation and Highways), 2010 SCC 4, [2010] 1 SCR 69, is an important decision of the Supreme Court of Canada about the interpretation and enforceability of exclusion clauses.

In this case [5], the Government of British Columbia ("BC") wanted to build a road. It issued a request for expressions of interest ("the RFEI") and received responses from six proponents, including Tercon Contractors Ltd. ("Tercon") and Brentwood Enterprises Ltd. ("Brentwood"). BC later issued a request for proposals for construction of the road ("the RFP"). The RFP explicitly stated that only the six proponents who had responded to the RFEI were eligible to submit tenders in response to the RFP. Because of this limitation on who could submit tenders, the RFP process deviated from a traditional public tender. Approval for this deviation was requested from and granted by the responsible minister.

Tercon submitted a bid in response to the RFP.

Brentwood was not in a position to submit a competitive bid alone and entered into a pre-bidding agreement with another company, Emil Anderson Construction Co. ("EAC"). The agreement stipulated that, if awarded the contract, Brentwood and EAC would enter into a joint venture agreement whereby they would share equally in the costs, expenses, losses, and gains of the project. Brentwood and EAC submitted a preliminary bid as a joint venture and a subsequent bid in Brentwood's name alone.

Tercon and Brentwood were the only two bidders to make the short list, and BC awarded the job to Brentwood. Subsequently, BC, Brentwood, and EAC took steps to hide the degree of EAC's involvement in the Brentwood bid. BC asked that EAC be recast as a Subcontractor, and the final contract entered into by BC was made solely with Brentwood.

Tercon sued BC for damages on the basis that BC had breached its contractual obligations to Tercon by accepting a bid from an ineligible bidder. BC denied liability to Tercon on two grounds. First, it argued that the Brentwood bid was not ineligible, and therefore BC had not breached its contractual obligations to Tercon. Second, it argued that even if it had accepted a bid from an ineligible bidder, any liability to Tercon was excluded by a provision of the RFP documents that read:

> *Except as expressly and specifically permitted in these Instructions to the Proponents, no Proponent shall have any claim for compensation of any kind whatsoever, as a result of participating in this RFP, and by submitting a Proposal, each Proponent shall be deemed to have agreed that it has no claim.*

In a 5-4 decision, the Supreme Court of Canada (SCC) found in favor of Tercon on both points. On the first point, the SCC held that BC knew that the Brentwood/EAC joint venture was an ineligible bidder, because it had not submitted a bid to the RFEI. By accepting the Brentwood bid, BC gave Brentwood an unfair advantage as compared to the other compliant bidders. BC's actions breached its contractual obligation to Tercon, as set out in the RFP. In addition, BC's actions breached the duty of good faith that it owed to the bidders.

On the second point, the SCC held that the language of the RFP did not exclude BC's liability to Tercon. The SCC provided within its analysis that effective tendering relies on the integrity and business efficacy of the tendering process. A key element of this process, especially in public procurements, is the equal treatment of bidders. The court held that only clear contractual language could exclude BC's liability for such an egregious breach of the foundational term of the RFP (i.e., the limit on eligible bidders) and of the duty of fairness owed by BC to all bidders. The SCC held that the language in the RFP was not sufficiently clear to achieve this end: when Tercon accepted the terms of the RFP by submitting a bid, it agreed to waive its right to seek compensation for those damages resulting from its participation in the bidding process. But, it did not waive its right to seek compensation for damages arising from the government taking steps that fell outside of the contemplated tender process, such as accepting a bid from an ineligible bidder. The SCC restored the trial award to Tercon of $3.5 million in damages and interest.

1.14 Privilege Clause

This term is widely used by a bid-calling authority to indicate that "the lowest or any tender may not be accepted." This type of privilege clause permits the Owner to decline acceptance of any tender based on reasonable and relevant grounds. Privilege grounds must be exercised fairly and based on objective reasons. A privilege clause should not be used to accept noncompliant bids.

If an Owner exhibits any preference in evaluating a tender or involves in bid shopping, he will be considered as having breached the terms of the bid contract in spite of the privilege clause. The Supreme Court of Canada in its decision (April 1999), M.J.B Enterprises Ltd. v. Defense Construction (1951), ruled that the privilege clause (lowest or any tender not necessarily accepted) does not permit the acceptance of a noncompliant bid

1.15 Rescission of the Contract

Contract rescission refers to the termination or cancellation of a contract. At any time prior to acceptance, the offer may be rescinded on notice. The purpose of contract rescission is to restore the parties to their original status before the contract was made. Contract rescission requires that all parties give back any benefits they have received while the contract was in force and be returned to their original states, as though the contract had never been formed in the first place.

A contract can be rescinded when one party uses false statements while entering into a contract; the misled party may give notice, prior to acceptance, or apply to a court to rescind the contract. The contract will be treated by a court as voidable, and it will be cancelled or set aside. However, the court may not grant a remedy if the injured party waited too long before claiming this remedy or was aware of misrepresentation and continued with the contract. The court will also determine for a valid basis to cancel the contract as the contract cannot be rescinded simply because the parties have changed their mind; both parties must show their clear intent to have the contract rescinded. In the event that only one party wants the contract to be rescinded, he must give proper written notice stating on what statutory grounds the rescission is requested, and it may be necessary to have a court of law determine whether a cancellation may be made.

1.16 Contra Proferentem

The principle of contra proferentem is a legal doctrine of contractual interpretation providing that any ambiguity in the Contract Documents will be interpreted against the party who prepared the document. This will only apply if the term is included unilaterally by one party and will not apply if the term was subject to negotiations by both parties. This even applies when the contract is in the standard form.

References and Further Reading

1. Course Notes. (2005). From engineering ethics and law attended by author – Conducted by EPIC Canada (Educational Program Innovation Center).
2. CCDC-2 (Canadian Construction Document Committee). (2008). Stipulated price contract.
3. Ashworth, A. (2006). *Contractual procedures in the construction industry* (5th ed.). England: Pearson Education Ltd..
4. Marston, D. L. (1981). *Law for professional engineers; Canadian and international perspectives* (3rd Edn). McGraw- Hill Ryerson Limited.
5. Carroll, D. P., & Lund, A. (2010, March). Case update: Tercon Contractors Ltd. v. British Columbia (Transportation and highways). E-Bulletin of Field Law's Construction Group. http://www.fieldlaw.com/publications
6. Moss, D., Buckley, G & Claire Rawlinson of Hammonds. (2008, March). Letters of intent – Contract formation, letters of intent, recent case law and practical considerations. Presented to ICES.
7. Ontario Construction Lien Act R.S.O. (1990). Chapter C30 – Last amended 2010 – Consolidation period: From July 1, 2011 to e-Laws currency Date Nov 26, 2013.
8. Notes from Principles of Construction Documents. (2014). Attended by the author – SAIT Polytechnic School of Construction, Calgary Alberta.
9. Naylor Group Inc. *v*. Ellis-Don Construction Ltd. (2001) 2 S.C.R. 943, 2001 SCC 58 (Supreme Court of Canada) – Lexum SCC.
10. Short, D. E.. Fasken Martineau DuMoulin LLP, Publication. The Construction Lien Act.
11. Samuels, B. M., & Sanders, D. R. (2011). Practical Law of Architecture, Engineering and Geoscience- Pearson Prentice Hall Canada.
12. Ontario Association of Architects. Lien hold is not for deficiencies or incomplete work. A Bulletin issued December 2012.
13. Bennett Jones LLP. Environmental Law – An article on doing business in Canada, December 2015.

Chapter 2
Parties to the Contract

2.1 Introduction

Construction work may include building of a new structure, additions, alterations, expansion, replacement, dismantling, erection, commissioning, rehabilitation, renovation, etc. Construction is a high-risk activity and must be actively managed from inception through completion.

When constructing a project, many parties/specialists are involved in the process of planning, designing, financing, monitoring, and building. Each of these parties has a different role to play, but they are temporarily joined together for certain period by a legal contract. However, this legal contract/agreement is signed by two main parties. The Owner, the first party, intends to carry out certain works for the implementation of a project and is sponsoring the works. The first party then appoints a Contractor, the second party, to execute the works.

No doubt the agreement is usually signed by two parties; however, taking a construction project from inception through to completion and commissioning requires the concerted and coordinated efforts of three distinct parties: the Owner of the project, the Consultant, and the Contractor. The Consultant is not a party to the contract but is appointed by the Owner to carry out duties that are stated in the contract.

The standard form of Contract FIDIC [1] identifies all three parties as "Employer," "Engineer," and "Contractor," whereas standard form of contract CCDC-2 [2] identifies all these three as "Owner," "Consultant," and "Contractor." Throughout this book, these terms wherever applicable will be used referring to FIDIC [1] and CCDC-2 forms; however, in general for the rest of the text "Owner" and "Consultant" terms will be used.

The goals and objectives of each of these parties are unique and often partially in conflict. In most situations the conflict and related disputes come to surface during the actual construction process. However, if the Owner understands the role and responsibilities of its construction Contractor, it will increase the prospects of a

A. Surahyo, *Understanding Construction Contracts*,
DOI 10.1007/978-3-319-66685-3_2

project with minimal conflict and adversity and greater opportunity for success. For most projects, the Contractor has little or no involvement until the bidding phase. During the planning and design phases, the focus is on the design of the project and in the preparation of the Contract Documents. It is the Contract Documents that contain the definitions of the roles and responsibilities of the Owner, the Consultant, and the Contractor during the construction phase.

2.2 The Owner

The Owner, when planning to construct a project, will have to face many challenges, such as time and cost constrains, program and quality goals, selection of a management team, a Constructor, etc. The Owner basically decides the scope, program, and budget for a project before starting the project design. During the design and construction phases, the Owner monitors the progress of work being done including its quality and makes periodic payments to the Consultant and Contractor.

The Owners within the construction industry could be from public sector as well as private organizations and may include:

1. Federal government.
2. Provincial, cities, and municipalities government departments.
3. Nationalized departments (autonomous bodies) like airports, hospitals, large industries, etc.
4. Private sector: it includes all those organizations that are not directly funded by some form of government money supply. The owner of a private sector may be a house owner, a builder, or multinational company.

In traditional contracts, the Owner has separate contracts with the Consultant and the Contractor. The Consultant provides professional services to undertake the responsibility of design and construction administration of the required project, whereas the Contractor is responsible for the construction phase. The success of every construction project depends as much on the project Owner as it does on the Consultant and Contractor.

As soon as the Consultant is appointed by the Owner, a sequence of activities like feasibility report, sketch design, detailed design, bill of quantities, Contract Documents, tendering, construction supervision, and commissioning are undertaken. All these activities involve many professionals like architect, structure and services engineers, quantity surveyor, geo-consultants, General Contractor, subcontractor, etc.

Generally the role of the Owner may include:

1. Identify needs.
2. Set out the parameters that define the project.
3. Establish reasonable goals for the project.
4. Arrange adequate funds.

5. Select appropriate consultants and appoint authorized representatives.
6. Develop and understand reasonable expectations for the project, its goals, and the parties involved in design and construction.
7. Negotiate the procurement package.
8. Arrange timely possession of site to the General Contractor.
9. Develop the understanding of responsibilities of the parties and their risks and providing authority to manage them.
10. Provide timely responses to the Consultant's submissions.
11. Provide timely responses to the Contractor's submittals, request for information, proposed changes, and claims forwarded by Consultant.
12. Issue timely progress payments.

Each Owner needs to define the project in terms of scope, standards, and outcomes and to manage the contracted relationships to ensure successful completion. The more complex the project, the more necessary it is for these organizations to have the capability, skills, and expertise, which are applied to procurement planning, design and documentation, tendering, and project and contract management.

2.2.1 Rights and Obligations of Owner

The contract prescribes the nature of the relationship. The Conditions of Contract are the primary source for determining the responsibilities and obligation of all parties. As per CCDC-2 [2] and FIDIC Conditions of Contract [1], some of the Owner's important responsibilities includes:

(i) CCDC-2 Sub-clause 3.2.2 provides that when Owner performs work for some other parts of the project either with his own forces or through another Contractor, the Owner shall provide coordination within the work activities among all the parties on site, assume overall responsibility of health and safety, take all reasonable precautions to avoid labor disputes or other disputes on the project, enter into the separate contracts with other Contractor compatible with the conditions of the contract for General Contractor, and ensure that insurance coverage is provided to the same requirements as of the insurance coverage of the General Contractor.

FIDIC Conditions of Contract [1] under Sub-clause 2.3 also calls for the Employer to make sure that Employer's personnel and other contractors working on site cooperate with Contractor's efforts under Sub-clause 4.8 (a), (b), and (c) for safety requirements and under clause 4.18 for protection of the environment.

(ii) CCDC-2 [2] Sub-clause 5.1.1 specifies that on request of the Contractor, before signing the contract, and from time to time, the Owner must provide the Contractor reasonable proof that financial arrangements are available to adequately fund the project. Under Sub-clause 5.1.2, the Owner has to give notice in writing to the Contractor if there is any material change to the Owner's financial arrangements.

FIDIC [1] Conditions of Contract include a similar provision in Sub-clause 2.4; however, FIDIC recommends a 28-day period to show evidence of financial arrangements after receiving request from the Contractor.

(iii) CCDC-2 [2] Sub-clause 5.3.1.3 provides that the Owner shall make progress payment to the Contractor on or before 20 calendar days after the later of receipt of the Consultant's application for payment or the last day of the monthly payment period. If the Consultant fails to issue a payment certificate or Owner fails to pay the Contractor due certified amounts or to provide evidence for availability of sufficient funds to meet the contract requirements, then in accordance with Sub-clause 7.2.3 and 7.2.4, the Contractor may give notice in writing to the Owner requiring a correction to the default within 5 working days. If the default is not corrected within those 5 days, the Contractor may suspend or terminate the contract.

Whereas, FIDIC Conditions of Contract [1], under Sub-clause 14.7(b), provide that the Employer shall make interim progress payments within 56 days after the Engineer receive the Contractor's statement and supporting documents. If the Owner fails to issue payment, then Sub-clause 16.1 provides that the Contractor may give 21-day notice and then suspend or reduce the rate of work. Additionally, under Sub-clause 16.2(c), the Contractor is entitled to terminate the contract if payment is not received within 42 days following the due date.

(iv) FIDIC Conditions of Contract [1] under Sub-clause 15.2 entitles the Employer to terminate the contract upon giving 14-day notice to the Contractor, if the Contractor fails to meet the requirements of performance security under clause 4.2 or abandons the work or have intentions not to continue with work; fails to proceed with work in accordance with clause 8 of commencement, delays, and suspension; or fails to comply with instructions issued under Sub-clause 7.5 for rejected plant, material, and workmanship or Sub-clause 7.6 for remedial work within specified period of 28 days or subcontracts the whole of the works without the required agreement.

The Employer is also entitled to terminate the contract immediately as per subparagraph (e) and (f) of the same FIDIC [1] Sub-clause 15.2, if the Contractor becomes bankrupt; or he or his personnel gives or offers any bribe, gift, gratuity, commission, or other thing of value.

The Employer can also terminate the contract at any time for his convenience under Sub-clause 15.5 without Contractor's default or any other justification.

Whereas, CCDC-2 [2] under Sub-clause 7.1.1 entitles the Owner to terminate the contract if the Contractor becomes bankrupt. Additionally under Sub-clause 7.1.2 and 7.1.4, the Owner may also terminate the contract in whole or part if the Contractor fails to correct a default in the time specified or the Owner may elect to correct such default and deduct the cost as certified by the Consultant from Contractor's due payments.

2.3 The Consultant

The Consultant in the context of a construction project means a consulting firm, the Consultant's authorized representative. The Consultant may be an architectural firm or an engineering firm or a sole proprietorship firm that is legally engaged/licensed in rendering professional services.

The role of the Consultant varies based on the services required. For full professional services, under a traditional approach, the Consultant is responsible for two separate functions: design and construction administration. Construction administration is essentially the monitoring of the Contractor's work in executing the design. The responsibilities incumbent on the Consultant as well as the limits of authority should be clearly defined in the agreement between the Owner and the Consultant and within the construction Contract Documents.

In general, the role of the Consultant, based on provision of full professional services, will cover stages of preconstructing, construction, and post-construction phases. Services provided by the Consultant as part of the preconstruction phase include:

1. Determining the feasibility of constructing a project.
2. Preparing preliminary budget and cost estimates.
3. Undertake conceptual, preliminary, and detailed design including site investigations.
4. Developing and evaluating alternatives with respect to design, location, and types of construction.
5. Preparing an environmental assessment and impact studies.
6. Assisting in obtaining approvals of authorities having jurisdiction over the project.
7. Preparation of Contract Documents.
8. Preparation of bid documents and contract formation.

The Consultant is responsible for the timely and impartial administration of the construction contract between the Owner and the Contractor. The services under this role of construction and preconstruction phases may be summarized by the following:

1. Direct and manage project execution.
2. Responding, advising, and consulting with the Owner during construction.
3. To ensure Contractor submits required insurance policies and performance bond prior to start work.
4. To issue information and instructions to the Contractor as the work proceeds.
5. To ensure that the work carried out conforms to the drawings and specifications, meeting with the required quality standards.
6. To watch for faulty workmanship or material incorporated in the work and issue instructions for remedial measures.
7. To monitor Contractor cost, schedule, and technical performance.
8. To ensure that the Contractor complies with all applicable safety regulations.

9. To liaise between Contractors and other authorities and organizations for the smooth completion of the project.
10. To verify the measurements of the work done by the Contractor and evaluate and certify contractor's applications for payment.
11. To review and approve Contractor's submittals such as work programs, shop drawings, product data and samples, etc.
12. Arrange monthly meetings to keep the owner well informed about the construction matters and to resolve Contractor's issues and other related site and work progress problems.
13. To prepare notices of change and change orders.
14. To review and evaluate Contractor's claims with respect to changes in contract sum or contract time and recommend to Owner.
15. To act as a channel for all claims and disputes between the Owner and Contractor and provide record and facts relevant to that.
16. To ensure that all records like files, tender drawings, construction drawings, and any superseded drawings are maintained properly.
17. To review of an application for substantial performance of work noting defects and deficiencies observed in the work.

2.3.1 Rights and Obligations of the Consultant

Some of the Consultant's important responsibilities as outlined in CCDC-2 [2] and FIDIC Conditions of Contract [1] are:

(i) FIDIC Conditions of Contract [1] provides under clause 3 that the Engineer shall be deemed to act for the Employer while carrying out his duties as per contract and shall have no authority to amend the contract. It further specifies that any approval, check, certificate, inspection, consent, instruction, examination, test, notice, proposal, request, or similar act by the Engineer shall not relieve the Contractor of his obligation to identify to the Owner any perceived errors, omissions, discrepancies, and non-compliance.

CCDC-2 [2] under Sub-clause 2.2.1 provides that the Consultant will administer the contract as described in the Contract Documents and clarifies under Sub-clause 2.2.6 that the Consultant will not be responsible for and will not have control, charge, or supervision of construction means and methods or safety precautions of the Contractor or for the acts or omissions of the Contractor, subcontractor, suppliers, or their employees or agents or any other persons performing portions of the work. Also the Consultant shall not be responsible for the Contractor's failure to perform work as required under contract.

(ii) FIDIC Conditions of Contract [1] provides that whenever required under contract that the Engineer shall proceed under clause 3.5 to render judgement on an issue (e.g., any claim dispute), the Engineer should consult both the parties to reach an agreement. This compels the Engineer to perform a role similar to a mediator. If no agreement can be reached, the Engineer has to make a fair determination in accordance with the contract, taking due regard of all relevant circumstances. However, as per clause 3.2, the Engineer cannot delegate this task unless otherwise agreed by both parties. If the dispute continues, then FIDIC [1] Sub-clause 20.2 requires that the parties shall jointly appoint a dispute adjudication board (DAB) by the date stated in the appendix to tender.

The DAB comprises one or three people (who are usually appointed at the start of the project) to render impartial decisions on any construction disputes. If a party is not satisfied with the DAB decision, the dispute can be referred to an arbitrator.

Whereas under Sub-clauses 2.2 of CCDC-2 [2], as further discussed in Chap. 17 regarding dispute resolution, it is specified that the differences between the parties to the contract on issues of work performance or interpretation of the Contract Documents shall be initially referred in writing to the Consultant, who will not show partiality to either party in evaluation of such interpretation and findings and reply his conclusion in writing to the parties within reasonable time. However, if the dispute is not resolved, then a procedure as set out in Sub-clauses 8.1.3 and 8.2.3 to 8.2.8 of negotiation, mediation, and arbitration shall apply to that dispute.

2.4 The Contractor

The Contractor is second party to the contract and, therefore, the Contractor is mentioned extensively in every Conditions of Contract. The Contractor under most forms of contract is required to execute and complete the works and remedy any defect in accordance with the Contract Documents and the instructions from the Consultant.

Despite various difficulties and inclement weather conditions or troubles, the Contractor is expected to complete the project and finish the work in the prescribed manner. The Contractor, who is responsible for constructing the entire project, is referred to as the "main," "prime," or "general" Contractor. Contractors may also function as a subcontractor or specialty Contractor, in which case will have responsibility for only a limited aspect of the project. All contractors will agree to do awarded work within a stipulated time period and for an agreed amount of money.

The Contractor's main responsibilities can be summarized as follows:

1. Supply and provide labor, material, and equipment as necessary to perform the works within the contract period.
2. Provide competent, experienced management and supervision for performance of the work.
3. Comply with all statutory laws and regulations during execution of the works and ensure that all who are employed on the site abide by these conditions.
4. Select and provide the means, methods, techniques, sequences, and procedures of construction.

5. Initiate, maintain, and supervise all safety precautions and programs for all personnel on site and the general public who may be affected by the works. In this regard, the Contractor shall ensure that all personnel on site are adequately trained and observe safe working procedures.
6. Implementation of quality control systems for aspects of the works as specified.
7. Schedule and coordinate the work with other groups like suppliers, subcontractors, etc. working on site.
8. Warrant and guarantee that all work done will be in accordance with the Contract Documents and will not be defective.

The cooperation and assistance of the Owner, Contractor, and Consultant is critical to the success of the project. Prior to starting works, the Contractor should submit any securities, guarantees, and insurance policies required by the contract. The Contractor is responsible for the security of the works throughout the construction period until the works are taken over by the Owner. The Contractor is required to provide insurance, not only to meet his direct liability but also for the Owner's protection. He is expected to exercise every reasonable safeguard for the protection of persons and property related to work under contract and adjacent to the construction site.

Building and civil engineering firms or companies can operate on a local, national, or international level and can be divided into small, medium, large, or very large concerns. It is desired that all parties engaged in construction must abide by some code of conduct. The Owner, Consultant, and Contractor must remember that they share a common goal, which is the successful completion of the project. The Construction Industry Development Board of the United States has published principles governing the conduct of parties. It says that in the interest of a healthy industry that delivers value to owners and society, the parties in any public or private construction-related procurement should abide by the following in their dealing with each other [3]:

1. Behave equitably, honestly, and transparently.
2. Discharge duties and obligations timeously and with integrity.
3. Comply with all applicable legislation and associated regulations.
4. Satisfy all relevant requirements established in procurement documents.
5. Avoid conflicts of interest.
6. Not maliciously or recklessly injure or attempt to injure the reputation of another party.

2.4.1 Rights and Obligations of the Contractor

Some of the Contractor's important responsibilities outlined in CCDC-2 [2] and FIDIC Conditions of Contract [1] are:

(i) CCDC-2 [2] under Sub-clause 3.1 provides that the Contractor shall have total control of the work and shall effectively direct and supervise the work ensuring that it conforms to the Contract Documents and shall remain solely responsible for construction means, methods, techniques, sequence, and procedures and for coordinating the various parts of the work.

However, it is clarified by CCDC-2 [2] within clause Sub-clause 5.9.1 that no payment under the contract or occupancy of work (partial or full) by the Owner shall constitute an acceptance of any portion of work or product which is not in accordance with the requirements of the Contract Documents.

(ii) Under Sub-clause 3.5 of CCDC-2 [2], the Contractor is required to prepare and submit before first application of payment, a construction schedule in conformity with the contract time, indicating the timing of the major activities of the work including with details of critical events. The schedule shall be updated on a monthly basis based on progress of the work. The Contractor shall advise the Consultant of any revisions to the schedule resultant from a request for extension of contract time.
(iii) CCDC-2 [2] under Sub-clause 3.7.1.3 provides that the Contractor shall be fully responsible for acts and omissions of subcontractors, suppliers, and persons directly or indirectly employed by them.
(iv) CCDC-2 [2] under Sub-clause 6.6.1 specifies that if the Contractor intends to make a claim for an increase to the contract price or if the Owner intends to make a claim for credit, the related party shall give timely notice in writing of intent to claim to the other party and Consultant.
(v) Under clause 7.2 of CCDC-2 [2], the Contractor is entitled to suspend the work or terminate the contract by giving such notice in the following events:

1. If the Owner turns bankrupt.
2. If the work is either delayed or suspended without the fault of the Contractor for a period of 20 working days or more by court order or other public authority.
3. If the Owner fails to provide reasonable financial arrangements on request of the Contractor.
4. If the Consultant fails to issue payment certificate in specified time or Owner fails to pay any due payments within specified time period.
5. If the Owner violates the contract requirements to a substantial degree and Consultant confirms in writing that the Contractor is entitled to give 5-working-day notice to the Owner and Consultant to rectify the default and if Owner fails to do so, the Contractor may suspend the work or terminate the contract.

Whereas FIDIC Conditions of Contract [1] addresses Contractor's entitlement for suspension of work and termination of contract under clause 16 and conditions mostly matches CCDC-2 [2] conditions discussed above.

(vi) Other Contractor's obligations under CCDC-2 [2] are: the Contractor shall be solely responsible for construction safety at the place of work (Sub-clause 9.4.1); and the Contractor shall provide all required insurance policies under clause GC 11.1, evidence of compliance with workers compensation legislation (Sub-clause 10.4), and required contract security (Sub-clause 11.2). The Contractor shall protect the work and Owner's property including property adjacent to the place of work from damage due to Contractor's operations (Sub-clause 9.1).

(vii) FIDIC Conditions of Contact [1] define most of the Contractor's obligations under clause 4. Some of the Contractor's main obligations covered under clause 4 are briefed as follows:

1. The Contractor shall execute and complete the works in accordance with the contract and with the Engineer's instructions and to remedy any defects in the works.
2. The Contractor shall be responsible for the adequacy, stability, and safety of all site operations and of all methods of construction.
3. The Contractor shall provide a performance security, in the amount stated in the Contract Documents, deliver to the Employer within 28 days after receiving the letter of acceptance, and shall send a copy to the Engineer,
4. The Contractor shall appoint the Contractor's representative with authority necessary to act on the Contractor's behalf.
5. The Contractor shall not subcontract the whole of the works and shall be responsible for the acts and defaults of his subcontractor, agents, or employees and comply with applicable safety regulations (also addressed under Sub-clause 6.7).
6. The Contractor shall set up a quality assurance system to meet the requirements of the contract. The Engineer shall be entitled to audit any aspect of the system.
7. If the Contractor encounters unforeseeable adverse physical conditions on site, the Contractor shall give notice to the Engineer as soon as practicable.
8. The Contractor shall take all reasonable steps to protect the environment (both on and off the site) and to minimize damage and nuisance to people and property resulting from pollution, noise, and his other operations.
9. The Contractor shall submit monthly progress reports and shall be responsible for keeping security on site.

Additionally, clause 8.7 of FIDIC Conditions of Contract [1] specifies that if the Contractor fails to complete work within the specified completion time period, the Contractor shall be liable to pay delay damages to the Owner. Sub-clause 13.1 provides that the Contractor is bound to carry out any variation/changes to the quantity, quality, and other characteristics of the items of work. These changes may include omissions, additions of work or services deemed necessary for permanent works, or changes to the sequence or timing of the execution of the works. The Contractor shall not carry out any addition or alteration unless an approved change order or variation is issued by the Engineer which is signed by the Owner.

For complete details with respect to Contractor's obligations FIDIC, [1] and CCDC-2 [2] can be referenced.

Reference and Further Reading

1. The FIDIC. (1999). Conditions of contract for construction – Red book; Author FIDIC.
2. CCDC-2. (2008). (Canadian Construction Document Committee) – Stipulated price contract.
3. CIDB (Construction Industry Development Board). Principals governing the conduct of parties, Brooklyn Square USA.

Chapter 3
Forms of Construction Organizations

3.1 Introduction

As a prospective business Owner, one of his first decisions would be choosing a business structure that will suit his needs. Knowledge of the construction business organizations operating within the construction industry is therefore essential for an Owner and Engineer. The main basic forms of construction organizations are as follows:

(a) The sole proprietorship
(b) The partnership
(c) The corporation
(d) The joint venture

3.2 The Sole Proprietorship

A sole proprietorship is the most common form of construction organization. It is the simplest and least expansive method for establishing a construction business as it is owned and operated by just one person. This form necessitates very few legal requirements to set the business which include that the name of company must be registered with the registrar of the business as well as appropriate tax authorities to get a license to work as a contractor.

In Canada, a business name registration refers to a registration under the "Business Names Act." It expires after 5 years and must be renewed. Ontario, Canada, assigns a business identification number (BIN) when a business name is registered in Ontario. It should not be confused with the Federal Business Number (BN) assigned by the Canada Revenue Agency (CRA) for federal programs, including goods and services tax/harmonized sales tax (GST/HST), import/export accounts, payroll deductions, and corporate income tax.

A. Surahyo, *Understanding Construction Contracts*,
DOI 10.1007/978-3-319-66685-3_3

The sole proprietor operates as his own boss and has complete control over his business. He has the freedom to make any and all decisions unilaterally. He keeps all of the profit alone and does not have to share with others. The sole proprietor has low set-up costs relative to those of a corporation. All business transactions and contracts are made in the name of his firm providing him the freedom of action. Benefits include that any business losses may be applied against other incomes of the proprietor.

Conversely, as being an individual proprietor, losses are born entirely by the one person. The Owner remains liable for all debts, obligations, and responsibilities of the business. The Owner is also at risk for personal liability incurred through acts of the Owner's agents or employees. The other disadvantage is that at certain income levels, a sole proprietor is taxed at higher rates than corporations. In the event of the death of the Owner, the business cannot be continued, unless he will nominate someone to continue the business.

3.3 The Partnership (General)

A partnership is a group or association of two or more persons, who agree to own and run a business together. All partners contribute to the capital of the business and shares in the management of the business. Profits or losses are usually shared in the same proportion as the distribution of ownership.

In a business partnership, an individual partner cannot sell or mortgage partnership assets. The partnership company, like individual proprietorship, will also be registered with designated public authority as well as income tax authority. Usually the tax is charged to individual partners separately based on the overall income. In Canada, partnerships are governed by the "Partnership Act" in each province.

Primarily partnerships are setup by written agreement called "partnership agreement" or "deed of partnership." The agreement must state clearly the rights, responsibilities, and obligations of each partner. All partners can mention into a deed any points which are mutually accepted to each other and which will safeguard against any misunderstanding arising at a later stage.

The advantages of partnership are that, more capital can be invested into the construction business, the responsibilities and losses will be shared. The disadvantages could be that the company does not have a separate legal identity. If one of the partners dies or becomes bankrupt or withdraws, the partnership will be dissolved, unless a provision in the agreement is made that the business will continue and that the remaining partners will purchase the shares of the departed partner.

Moreover, in the event of disagreement, consulting all partners for decision will delay the construction and related issues. If one partner is inefficient or dishonest, the other partners will suffer due to his wrong actions. Each partner is jointly and severally liable for company debts. For example, if the partnership assets are insufficient to satisfy a creditor's claim, the partners' personal assets are subject to attachment and liquidation to pay the business debts.

Each general partner is deemed the agent of the partnership. Therefore, if any partner is carrying on partnership business, all general partners will be held liable for that partner's dealings with third persons.

3.3.1 *Limited Partnership*

In an ordinary or general partnership discussed above, each partner has unlimited liability to creditors. However, some partners may be willing to invest money in the firm but do not wish to participate in its management and have no voice or vote in the matters of management. Such a partner is called a limited partner, and the agreement is termed as a "limited partnership" also called sleeping partners.

Unlike the general partner, the limited partner is liable for partnership debts only to the amount of his investment in the partnership. The number of such partners could be one or more in the company; however, at least one of the partners must be a general partner, who could run the business. The limited partnership is usually used as a means of raising capital for the company. In case of death, disability, or withdrawal of a general partner, the partnership will be dissolved, unless the partnership agreement provides or all partners agree, in writing, to substitute a general partner. However, death or incompetence of a limited partner has no effect on the partnership.

3.4 Corporations

A major difference between a corporation and partnership is that a corporation is a legal entity that is separate from its shareholders/owners. Shareholders may receive dividends from the corporation and are entitled to receive the assets of the corporation upon liquidation once the corporation's debts and other obligations have been discharged. A shareholder of a corporation is not personally liable for the debts, obligations, or acts of the corporation. Whereas, a partnership is a business entity with individuals who share the risk and benefits of business. A corporation is a separate legal unit from its owners which means a partnership exists separately from the Owners/shareholders and will continue to exist if one of the owners should die or leave the business or if the ownership of the business changes. A corporation can make contracts or legal agreements, and its accounts are kept separate from the accounts of the owners. Like sole proprietorship and partnership, corporations also need to be registered in every province of Canada in which it does business as well as federally with Canada Revenue Agency.

Corporations are jointly owned by the people who have invested in the business. These people buy shares in the corporation, and they are therefore called shareholders. These shareholders appoint Directors to run the business. In a private limited

(Pvt. Ltd) company, the Directors are usually the most important or majority shareholders.

The advantages of private limited corporations are that the shares can be sold to a large number of people, usually relatives and friends, as they cannot advertise to sale the shares to the general public. The sale of shares leads to much larger sums of capital to invest in the business.

All shareholders have limited liability. It means that if the corporation fails with debts owing to creditors, the shareholders could not be forced to sell their possessions to settle the debts. The shareholders will only lose their original investment in the shares. Their liability is limited to that original investment. This is a major benefit compared to the position sole traders, and partners can find themselves in, if their business fails.

Since a corporation is treated as a separate legal unit under the law, it has the same rights as an individual, in that it can own property, investments, carry on business, borrow money, incur liabilities and debt, sue and be sued, etc. The corporation, being an independent unit, must have its own separate bank account, and the corporation's assets should not be mixed with the personal assets of its owners. The corporation must maintain its own set of financial/accounting books or records and must file corporate tax returns that are separate from the personal income tax returns of its owners.

The disadvantages are that incorporation is more complicated, expensive, and time-consuming to establish and maintain. Because corporations are regulated and monitored by the government, there is additional paperwork involved in their maintenance. In addition to extra bookkeeping, a minute book containing corporate bylaws, meeting minutes or resolutions, registers, and ledgers must be maintained.

Each individual in the corporation will be required to file two or three tax returns each year: one for their personal income and a separate one or two for their corporation's income. This will mean extra accounting fees and more time in bookkeeping. There is also possibility of higher taxation for incorporated businesses by double taxation. Corporate income is taxed at the corporate level when it is earned by the corporation, and it is taxed again at the personal level when it is removed from the corporation as a dividend. This effectively means that this income will be taxed twice. When losses arise in a corporation, they can only be offset against the earnings in that corporation; they cannot be transferred to the shareholders [1].

The other disadvantages are that a private limited corporation cannot be sold or transferred to anyone else without the agreement of the other shareholders. The account of a corporation does not remain as a secret; each year the latest accounts must be sent to the registrar of companies. The corporation cannot offer its shares to the general public so as to raise really large sums of capital to invest or expand the business.

Most of the engineering corporations start business as private companies. However, some corporations then may decide to convert to public corporations. In that case, the shares can be sold to the general public. Corporations that are controlled by the same person or group of persons are called "associated corporations."

3.5 The Joint Venture (JV)

For large scale projects, it has become popular for two or more contracting firms to unite forces through the setup of a joint venture. The members of a joint venture may be sole proprietorships, partnerships, or corporations, but the joint venture itself is a separate business entity and essentially a partnership limited to one particular project.

A joint venture allows the partners to combine resources, assets, and skills. This results in the advantage of pooling construction tools and plants, office facilities, personnel, and financial means. Each joint venture is responsible for the conduct of the work and shares in any profits or losses.

Joint ventures are frequently between local and foreign companies. Each foreign market offers unique opportunities and risks, and many firms naturally look to JVs with one or more partners for assistance in entering new markets. JVs have become a major feature of the international business landscape due to increased global competitiveness and technological innovation.

In Canada, there are no federal or provincial requirements that a foreign Designer or Contractor enter into a joint venture with a local contractor in order to design, build and be paid for their work. However, a foreign Contractor may be required by provincial law to register or be licensed as an "extra-provincial" corporation and may also require provincial or municipal registrations or licenses, or both, to engage in certain regulated activities. Foreign designers must be licensed by the Canadian province or territory in which they intend to work [2].

The formation of a JV can be a complex process. After a compatible joint venture partnership (JVP) is selected, the specific goals of the enterprise must be defined. The JV agreement should define clearly the scope of venture, obligation of the parties, working capital, percentage interest of each partner, division of profits and losses, etc.

3.5.1 Basic Elements of a Joint Venture

Basic elements constituting a joint venture are:

1. The parties should enter into a contract or an agreement that specify their mutual responsibilities and goals.
2. The parties should have joint control and management over the venture.
3. The parties should have a share in both loss and profit.
4. The parties should contribute property, cash and organizational capital, skills, knowledge, or efforts to achieve common goals.
5. The parties should work in a good faith in matters that concern the common interest of the venture.

It is not necessary that all items described above must be presented in establishing a joint venture. Simply a joint venture depends upon three elements: joint ownership, joint operation, and an express or implied agreement [3].

Moreover, the elements are required to establish JV matches with those of partnership but are not the same. A JV is more limited in scope and duration. Additionally, a JV is related to a single transaction, means for a particular purpose or project, while a partnership is typically related to a general and continuing business. Additionally, in a joint venture, each participant retains separate ownership of their property and shares only the expenses and revenue of the particular project or venture.

Another important difference between, a joint venture, partnerships and corporations is the manner in which courts assign liability when problems arise on a project. For partnerships, the law is quite clear; each partner is generally responsible for the entirety of all debts and liabilities of the partnership. There are exceptions to this general rule as some partnerships are designed specifically to limit the liability of individual partners. Nonetheless, full liability for each partner is the default rule. In joint ventures, by contrast, the liability of individual members is less certain. There is no simple default rule for liability. A clearly worded joint venture agreement can, however, help to determine how a court will apportion any liability between the members of the joint venture.

For corporations, the law is also quite clear; the corporation has a distinct legal personality, and it alone is liable for its debts, rather than its shareholders and officers [4].

3.5.2 Termination of Joint Ventures

A JV may be terminated due to a breach of the joint venture agreement. A JV may terminate upon achieving its objectives at a time specified in the contract as well as upon failing to meet its objectives. In addition, a change in the JV's objectives or those of a shareholder may also lead to the early termination of the JV. Disagreement by JV partners on fundamental management issues may also lead to termination. A JV may also be terminated due to sudden death of an active member or if a court decides termination is warranted given serious disagreements between the members making its continuation impractical.

References and Further Reading

1. Cohen, E. P. Barrister & Solicitor – New business Now.com, a division of NBN Business Services Inc. Selecting a business structure in Canada.
2. Reynolds, B., Vogel, S., & Houle, Y. Borden Ladner Gervais LLP Canada, getting the deal through – Construction 2009.
3. Uslegal.com – An article on elements of joint venture relationship.
4. Berezowskyj, S., & Daigle, M. Singleton Urquhart, joint venture – Disjointed Liability, BC, Canada.

Chapter 4
Contract Delivery Methods

4.1 Introduction

Once the Owner decides to proceed with any project, he has to determine which contract delivery method is suitable to achieve the primary objectives of the project. Specifically, to get it constructed on time, within budget, to the desired quality and with minimum exposure to risk. The selection and use of an appropriate contract procurement/delivery method are fundamental to the success of a construction process.

A project delivery method is a system used by an agency or the Owner for organizing and financing design, construction, and all related services for a facility through an agreement with one or more parties. The contract method selection involves the analysis of many issues such as cost certainty, time certainty, speed, flexibility, risk involvement, etc. Each contract delivery system has its advantages and disadvantages. Each construction method represents a particular allocation of risks in exchange for anticipated benefits. Hence, for choosing the right construction method, the Owner must be prepared to recognize that no method will be free of risk, so he should get information about possible risks for each construction method and plan ahead so as to minimize the risks of the selected construction method. An inadequate project delivery system often encourages adversarial relationship among the parties working on the project, increases the number of claims, and disputes and prevents the free flow of information necessary for the successful completion of the project.

Moreover, the objectives of a project such as quality, schedule, cost, and risk must be established prior to the evaluation of the available project delivery methods. The success of any of these methods is also very dependent upon the Owner's capability for managing the project delivery process. Thus, the Owner's involvement throughout the various phases of the project is important because the ultimate test of a successful project is the degree of Owner's satisfaction.

A. Surahyo, *Understanding Construction Contracts*,
DOI 10.1007/978-3-319-66685-3_4

4.2 Contract Delivery Methods

Owners and projects have different needs, such as urgent time frames, funding pressures, increased safety and quality requirements, etc. Once the project objectives are established and the Owner's capabilities are defined, the role of the construction professional is to identify Owner's needs and propose a means to deliver a completed project that satisfies both the needs of the Owner and objectives of the project.

Presently there are many procurement trends in the construction industry. The three primary and most commonly employed concepts are design-bid-build (traditional method), design and build, and public-private partnership with various variations.

4.2.1 Design-Bid-Build Method

This is the most popular and widely used construction method and is also known as the traditional method. With this method, construction is separated from design; hence, the Contractor has no responsibility for design. The Owner/Client contracts with a consulting firm, who designs the project, prepares detailed drawings, specifications, and bill of quantities with the coordination of various specialists. The bid documents are prepared and the project goes out for bid, and the contract is usually awarded to the lowest bidder. Thus, it gives rise to two different agreements: one between the Owner and Consultant and the other between the Owner and the Contractor.

Sometimes there are prequalified (short listed) bidders, and sometimes the bidding is open to all contractors interested in bidding. The Contractor executes an agreement with the Owner to produce the project in accordance with those plans within the stipulated time and for a fixed price or on unit price basis. The Contractor then manages the construction with his team and with the help of subcontractors and suppliers. During construction, the site supervisory staff of the Consultant visits (sometimes reside on site) and inspects the work to ensure that it is being completed in accordance with the drawings and specifications.

4.2.1.1 Strengths and Weaknesses

Strengths

1. Main advantage of this method is familiarity and experience with the many standard form Contract Documents typically used for such projects.
2. Owners have a good idea of final cost as the complete design process is almost finished before bidding.

3. Provide consultants better control of document preparation and the selection of contractors.
4. Offers the Owner/Consultant a significant amount of control over the end product because the project features are fully determined and specified prior to selection of the Contractor.
5. Due to competition it provides the lowest possible price and avoids favoritism.
6. It provides equal opportunity to all qualified bidders to obtain the construction contract.

Weaknesses

1. Greater coordination and control is required because several parties with different contractual relationships are involved.
2. Owner is in contract with multiple parties which can be a serious weakness if major defect arises.
3. Contractor has no input during the design phase and will not be able to propose innovative methods.
4. No pricing until bid phase is complete.
5. Project duration is longer as complete plans and specifications must be prepared before Contractors put a price to the project.

4.2.2 The Design and Build Method (DB)

Under this method, the Owner awards an entire project to a single entity, who takes overall responsibility for both the design and construction of the project. The Owner establishes project requirements by issuing an Owner's statement of requirements. The design-build arrangement combines the services of the Consultant/Designer and the General Contractor. The single source responsibility of the design-builder reduces administration claims and litigation conflicts between the design and construction teams as the division of responsibility between them is primarily the responsibility of the design-builder. By working together as a team, the Consultant and Contractor can fast track the process and improve constructability, quality, and innovation potential. All these lead to low cost for the Owner.

Usually there are two general organizational formats for design-builders. First, the Owner may contract with an integrated design-build firm. The integrated design-build firm has Architects, Engineers, and construction professionals, all in-house. The second design-build organizational format is the design-build joint venture. Under this organizational structure, the Designer and the General Contractor enter into a joint venture agreement for the duration of the project.

4.2.2.1 Strengths and Weaknesses

Strengths

1. Reduces the number of key contract parties from three to two.
2. Unlike traditional contracting, avoids adverse interest conflicts between the General Contractor and the Designer as the same party provides design and construction services.
3. General Contractor's claims for design errors are eliminated.
4. Another benefit is that the design-build approach allows construction projects to be completed in a shorter time frame (as construction can be commenced prior to full design).
5. Moreover, by having sole responsibility by one entity, the Owner can look to one party only for responsibility, should problems arise in the design or construction of the project.
6. With a design-builder involved in early stages, costs are calculated much earlier in the process.
7. Contractor input during design and planning phase is available.

Weaknesses

1. Can reduce the safeguards of a good check and balance system with the Designer and Contractor working together.
2. Reduced Owner control and participation. Frequently, the Owner will appoint an independent quantity surveyor or a Project Manager to keep a check on quality and cost.
3. Due to focus on optimum schedule, people are often rushed and make mistakes leading to problems which are difficult and expensive to resolve.
4. As one party is responsible for design and construction, design changes are made primarily for the benefit of the Constructor.
5. Finished product may be disappointing or unsatisfactory, if early plans and specifications are not clearly outlined by the Owner.

4.2.3 Other Forms of Design Build Concept

4.2.3.1 The Partnering (Public-Private Partnership (P3))

P3 is a contractual arrangement between two parties mainly a public authority and a private sector for either a specific period of time or an indefinite/long-term period. P3 is not a new form of contract – it is a procedure for making relationships work better. The arrangement features teamwork, continuous improvement, openness and acceptance of new ideas, trust, and mutual benefit. P3s are a long-term performance-based approach for procuring public infrastructure. The private sector assumes a major share of the responsibility in terms of risk and financing for the delivery and performance of the infrastructure.

The Canadian Council for public-private partnerships defines P3 as "a cooperative venture between the public and private sectors, built on the expertise of each partner that best meets clearly defined public needs through the appropriate allocation of resources, risks and rewards [1]."

The essence of a P3 arrangement is the sharing of risks. Central to any successful public-private partnership initiative is the identification of risk associated with each component of the project and the allocation of that risk factor either to the public sector or the private sector or perhaps a sharing by both. Thus, the desired balance to ensure best value is based on an allocation of risk factors to the participants who are best able to manage those risks and thus minimize costs while improving performance. Partners have to accept that problems will occur, so at the outset there should be an agreed procedure for dealing with all problems as early as possible, before they become disputes.

Many projects in Canada have been completed or awarded on the P3 model, some of them are 407 ETR toll highway in Toronto; Confederation Bridge linking Prince Edward Island to New Brunswick; water projects in Dartmouth, Winnipeg, Moncton, and Regional Municipality of York; Hamilton-Wentworth Airport; and the William Osler Health Centre Group comprising three hospital sites: Brampton, Etobicoke, and Georgetown [2].

In construction, there are many variations of P3s, but generally the setup is: the private entity will be comprised of a design-build team, a maintenance firm, and a landing firm. The most common models of P3s are:

- *Design-build-maintain (DBM)*: The private sectors design and construct the facility. The public entity takes ownership after construction is complete. The private sector continues to maintain the facility based on long-term maintenance agreement.
- *Design-build-finance-maintain (DBFM)*: In addition to DBM outlined above, sometimes the private sector also takes responsibility for the finances in the DBFM model.
- *Design-build-finance-maintain-operate (DBFMO)*: In addition to DBFM defined above, the private sector provides hard and/or soft facility management services as well as operations under a long-term agreement.
- *Build-own-operate (BOO)*: In this model, the private sector finances, constructs, owns and operates a facility or service permanently. The Public constraints are stated in the original agreement and through ongoing regulatory authority. Although the government does not provide direct funding in this model, it may offer other benefits such as tax exemption. This approach is quite common in the power generation sector.
- *Build-operate-transfer (BOT)*: This method almost matches with BOO model. Under this arrangement, the private sector designs and builds the project to the specifications agreed to by the public agency (host government and the Owner),

operates the facility for a specified time period, and then transfers the facility to the host government after the expiry period. In most cases, the private partner will also provide some or all of the financing for the facility, so the length of the contract must be sufficient to enable the private partner to realize a reasonable return on its investment through user charges. There are many examples of this type of project such as tunnels (Sydney Harbour Tunnel, Australia), bridges (Dartford Bridge, England), power generation stations (Shajiao Power Plant, China), etc.

4.2.3.2 Strengths and Weaknesses

Strengths

1. Uses benefit of private sector expertise in various fields
2. Transfers majority of risk share to private sector
3. Single point of responsibility
4. Uses funding sources which may not be possible through public methods
5. Long-term operations and maintenance

Weaknesses

1. Handling a P3 project is very complicated and an Owner will need a high level of expertise.
2. The Owner may experience higher total life cycle costs.
3. Potential for a lack of communication between the parties.
4. Difficult for small consulting firms and construction companies to compete in this market.
5. "Best value" is mostly not achieved.

The benefits of the combined inputs of project teams will be much more when partnering is initiated before the final design phase. In this way, it will result in cost reduction through an improved design and tailored construction methods on project delivery systems.

4.2.4 The Construction Management Method

The term Construction Manager (CM) basically refers to the professional construction management firm that is responsible for managing the entire construction process. Construction management is most beneficial when working on large, complex, and long-term projects where there are likely to be multiple Prime Contractors and the Owner requires extra management resources to control the project. There are many forms of management methods. The role of the CM in a project may vary substantially and can be performed under a variety of contractual terms. However, there are two commonly used approaches to construction management: (1) Construction Manager as Agent and (2) Construction Manager at Risk.

1. ***Construction Manager as Agent***

Under this form of construction management, the Owner contracts with a professional construction management firm who acts as the Owner's agent in the role of a Consultant. The CM is retained prior to the design phase providing preconstruction services such as value engineering, estimating, cost control, schedule analysis and bid review, and undertaking the administrative responsibilities of managing all trade contractors during construction. The CM will not perform himself any particular trade work or other construction works.

Under this arrangement, the Owner contracts directly with the required trade Contractors like a General Contractor to complete the project. Thus there will be three sets of contracts:

1. Owner – CM
2. Owner – Designer
3. Owner – Trade Contractors

The CM is retained on a fee for service basis and acts on Owner's behalf in managing and coordinating the trade contracts in the best interest of the Owner. The concept of this method is to use a professional on-site Manager whose sole function is to guide the project to an efficient completion. A major benefit to the construction management approach is that, under the disciplined scheduling of the Construction Manager, the project may be phased or fast-tracked whereby the design and construction periods are overlapped to permit an earlier start and completion of the construction.

2. ***Construction Manager at Risk***

Under this form of construction management, the CM is responsible for both services and construction. Under this arrangement, the CM contracts with all the trade Contractors to complete the project. With this method, there are two sets of Contracts by the Owner:

1. Owner – CM
2. Owner – Designer

The CM assumes responsibility for the performance of the trade contracts like a General Contractor under the traditional method and is paid for the work done by trade contractors either on cost reimbursement basis or fixed price basis. The CM is retained prior or during the design phase providing preconstruction and construction services. Under this concept, the CM holds the risk of subletting the construction work to trade subcontractors and guaranteeing completion of the project for a negotiated price following completion of the design.

4.2.4.1 Strengths and Weaknesses

Strengths

1. Issues of constructability, cost, and schedule are addressed during design phase: hence, cost savings can be made due to earlier start and prepurchasing.
2. Allows fast-tracking or phased construction, providing earlier completion.
3. A system of checks and balances exists.
4. No additional Owner's personnel are required to monitor construction.
5. Reduces variations and claims due to design errors.
6. The CM-as-Agent provides the Owner with maximum control over various aspects of the project having direct access to each team member.
7. The CM-at-Risk serves as a single point of responsibility contracting directly with the sub-trades during construction.

Weaknesses

1. The Owner does not know the overall bid price at commencement of construction.
2. Construction usually starts before completion of design, which may lead to variations; however, it could be minimized by an expert CM team.
3. The Owner bears additional cost of the CM fee.
4. Under the CM-as-Agent arrangement, overall schedule and cost risk remains with Owner.
5. For CM-as-Agent, no single source of contractual construction responsibility exists since all contracts are held by the Owner. The CM, not being the Contractor, is not responsible for construction means, methods, techniques, sequences, or procedures.
6. In CM-at-Risk, the CM takes responsibility for quality of work, costs, and completion date.

4.2.5 Single and Multiple Prime Contract Technique

Single contract is the most commonly used contract type. Plans and specifications are prepared by the Consultant and become part of the bidding documents. A single Contractor is then selected by the Owner to perform the work. A single contract is usually the easiest to administer because of its centralization of responsibility, that is, one Owner, one Contractor, and one construction contract.

With multiple prime contracts, the Owner divides a project into two or more parts and then enters into a separate contract for each part. The most frequent use of multiple prime contracts is for phased construction, in which contracts are awarded sequentially for each phase. This type of construction is also referred to as the "fast-track" method. Contracts for parts of the project, such as site development, site excavation, or foundation works like piling, etc., are awarded before the contract for the main structural work is awarded.

Sometimes, due to the size of a project, the Owner may divide the prime project into subprojects based on different trades or crafts like mechanical, electrical, HVAC, structural/civil, etc. and award separate prime contracts for each subproject. The related Prime Contractor is responsible for all work within its particular subproject.

Construction by multiple contracts requires significant coordination to define the scope of each contract package and the interaction of Contractors. It is best implemented by engaging an independent design professional and employing a Construction Manager (as discussed under construction management methods) who will work with the Designer to establish and coordinate scope of work for the various packages. For efficient management, the CM should be involved in the project from inception to completion.

With fast-track projects, it is beneficial to use a cost plus fee type of contract for the general construction. A fixed price agreement covering the total price for work may not be cost effective to the Owner since the CM would need to set the price of the project high enough to protect himself against the unknown cost which is inherent in assuming a fixed sum contractual obligation prior to completion of designs for the construction package. Alternatively, if the fixed price were set too low, the Owner must expect to be burdened with renegotiated agreements or the contentions which accompany numerous change directives and change order requests.

Multiple contract arrangement should be selected only after full consideration of all factors involved, including administrative services of a CM to coordinate bidding and site work.

4.2.5.1 Strengths and Weaknesses

Strengths

1. The major advantage is the time saving and early project delivery because the project is being built while the design and document preparations are being accomplished.
2. To expedite project delivery, critical equipment or materials may be ordered early in the process.
3. Early project completion reduces the Owner's interim financing costs.
4. Owner has financial benefit of early occupancy of the project site.
5. Early selection of the CM in the design phase can contribute to cost control on the project during document preparation.

Weaknesses

1. Final costs are unknown.
2. Early start of the construction phase prior to completion of overall design can lead to many changes in scope of work, resulting in delays and additional costs.
3. Multiple contract construction poses unique problems which require a particular expertise.
4. Litigation risks are high using incomplete Contract Documents.
5. Due to various contract packages, there are chances of omissions and duplications within the sets of drawings and specifications which will increase exposures to claims.

References and Further Reading

1. Canadian Council for Public- Private Partnerships. (2005). Definitions.
2. Donald G. Pierce et al (2003). Goodmans LLP. An article on "*The latest in delivery methods in Canada*".

Chapter 5
Types of Construction Contracts

5.1 Introduction

In general, the construction contract price includes direct and indirect project costs plus profit. While construction contracts serve as a means of pricing overall construction, they also structure the allocation of risk to the various parties involved. The Owner has to decide what type of contract should be used for a specific project to be constructed and to set forth the terms in a contractual agreement. It is important to understand the Owner and Contractor's risk profiles associated with different types of construction contracts.

There are several types of contracts based on pricing arrangements according to which Contractors are paid for their work done. There are also many variations and combinations that can be used; however, the most commonly used types and criteria are discussed here:

1. *Price based*: Lump sum or unit price contracts, in which the prices are quoted by the Contractor at bidding stage.
2. *Cost based*: Cost reimbursable contracts, in which the actual costs incurred by the Contractor are reimbursed together with a fee to cover overheads and profit.

5.2 Price-Based Contracts

5.2.1 Lump Sum Contract

A lump sum contract, also called stipulated price or fixed price contract, is the most basic form of agreement between Owner and the Contractor. In lump sum contracts, the Contractor calculates his rates based on the drawings and specifications

A. Surahyo, *Understanding Construction Contracts*,
DOI 10.1007/978-3-319-66685-3_5

prepared by the Designer. He then submits one lump sum price for the whole works or gives breakdown of the total sum against major activities or sections of the work. Payments made to the Contractor are usually assessed on the basis of percentage of work completed on a monthly basis. This is governed by CCDC 2 Conditions of Contract [1].

The lump sum contract is mostly used in projects or purchases where the plans and specifications are complete in detail before requesting bids. This procedure allows the Owner to know the cost of the project or purchase in advance. This type of contract is not advisable when plans and specifications are incomplete because the resulting bids will generally be inflated relative to actual cost as a reflection of the lack of certainty. To avoid changes, the scope of work must be well defined at the bidding stage.

Sometimes, lump sum contracts include the usual bill of quantities with estimated quantities, the rates against which are quoted by the contractors and total figure is treated as a lump sum price. In this case, a condition or provision shall be added in the contract saying "the approximate quantities are solely included for the Contractor's information. They will not be subject to re-measurement and no adjustment shall be made to the lump sum price in the event of the actual quantities of work executed by the Contractor differing from those included in Bill of Quantities except where such difference is the result of a change in scope." Also the rates quoted by the Contractor within the bill of quantities can be used as a basis for valuation of scope changes.

Basically, this type of contract is considered as a fixed price contract but seldom remains fixed during construction. Changes in scope, flaws in drawings and specifications, and changed site conditions cause variation to the original quoted contract price. Hence owners usually add contingency amounts in their estimates to address such issues.

5.2.1.1 Strengths and Weaknesses

In a lump sum contract, the Owner has assigned most of the risks to the Contractor. The construction means, methods, techniques, sequences, and procedures are the Contractor's responsibility. The Contractor, in turn, can be expected to ask for a higher markup in order to take care of unforeseen contingencies. Besides the fixed lump sum price, other commitments are often made by the Contractor in the form of submittals, such as a specific schedule, the management reporting system, or a quality control program. If the actual cost of the project is under estimated, it will reduce the Contractor's profit by that amount, whereas overestimate will have an opposite effect. Any modifications to scope, design, or schedule will give rise to a cost change. Contractors can bury anticipated or incurred costs within change orders during the construction process.

5.2.2 Unit Price Contract

A unit price contract is the traditional system and most popular in both the building and civil engineering sectors. Under this form, the detailed bill of quantities is prepared by the quantity surveyor of the Consultant based on drawings and specifications. The contractors quote their rates against these calculated quantities on basis of unit rate method. However, the contractors are paid for the work measured in place on the basis of actual quantity multiplied by quoted rate.

This method is advantageous when the quantities cannot be accurately identified in advance or in such works having a high content of ground surface work where the quantities are rather unpredictable. This method is particularly favored for linear construction like roads, railway tracks, sewers, water mains, or buried/aerial utility lines. The estimated quantities at the proposed unit prices submitted by bidders are used in comparing the bids. If changes occur, the unit prices and rates in the bill of quantities can be used as a basis for valuation. Typically, a change or variation in quantities up to 15% is permitted in this type of contract without the need of a formal change order as long as the items remain generally the same as in the initial contract. This is governed by CCDC 4 Conditions of Contract [2].

Note, as per FIDIC [3] Conditions of Contract, Sub-clause 12.3, if the actual measured quantities of the items vary more than 10% from the quantity of bill of quantities (schedule of rates), then a new rate shall be appropriate for an item. Whereas, according to the Contract Document CCDC-4 [2], if the actual quantities exceed or fall short of estimated quantities by more than 15%, then the Owner or the Contractor may request an adjustment to a unit price of the related item provided in the schedule of prices.

5.2.2.1 Strengths and Weaknesses

In this type, the Owner takes the risk of changes in the quantities originally estimated. The final cost of the project is not known to the Owner until completion of the project. Bidding can be called based on completed drawings only to start project early. Additional staff will be required to measure, control, and report on quantities. Changes in contract can be made easily as the Contractor will be paid against actual work done on site. The cost of tender is actually reduced because the bidders do not need to set up their own bill of quantities. If changes in scope of work occur, the unit prices and rates in the bill of quantities can be used as a basis for valuation.

Another type of contract available is a combination of lump sum and unit price contracts. There are advantages when a definite number of items, like superstructure of a building or a bridge, can be covered by the lump sum feature, and an indefinite quantity of items, like substructure of a building, can be included in the unit price method. This is common in many government projects.

5.3 Cost Reimbursable Contracts

Also known as cost plus contract, this type entails Contractor reimbursement of the actual cost of carrying out the work plus an additional amount in the form of fees to cover his overheads and profit. The Owner selects a Contractor recognized for dependability, experience, and skills through direct negotiation and establishes the terms of agreement between them and the amount of fee to be paid. The selected Contractor then starts the work and ensures all costs are transparent to the Owner including payments made for labor, material, machinery and equipment, subcontractors, etc. The Owner usually makes interim payments to the Contractor on a monthly basis. By proceeding in this manner, the Contractor gets all expenditures compensated along with his fee, which was determined at the outset of the project.

Under a cost plus contract, the Owner is responsible for payment of any costs resulting from unforeseen conditions. The scope of work also needs to be carefully defined to avoid any disputes.

The Contractor's fee is calculated in various ways such as:

(i) Cost plus percentage fee
(ii) Cost plus fixed fee
(iii) Cost plus variable fee
(iv) Guaranteed maximum price or target price contract
(v) Time and material

5.3.1 Cost Plus Percentage Fee Contract

With this form, the Owner assumes all risks of cost overruns. The Contractor will receive the actual direct job cost plus a fixed percentage of the construction cost and have little incentive to reduce job cost. Furthermore, if there are pressing needs to complete the project, overtime payments to workers are common and will further increase the job cost. Unless there are compelling reasons, like need for urgency, this method is not advisable to the Owner. This method can be used to reduce the time it takes to procure a Contractor.

5.3.2 Cost Plus Fixed Fee Contract

Under this type of contract, the Contractor will receive the actual direct job cost plus a fixed fee and will have some incentive to complete the job quickly since its fee is fixed regardless of the duration of the project. However, the Owner still assumes the risks of direct job cost overrun, while the Contractor may risk loss of profit if the project is delayed beyond the expected completion time.

5.3.3 *Cost Plus Variable Fee Contract*

For this type of contract, the Contractor agrees to a penalty if the actual cost exceeds the estimated job cost or a reward if the actual cost is below the estimated job cost. The Contractor's fee is made up of two parts: a fixed amount and a variable amount depending upon the relationship between the target cost and the actual cost. Furthermore, the project duration is usually specified and the contractor must abide by the deadline for completion.

This type of contract allocates considerable risk for cost overruns to the Owner but also provides incentives to contractors to reduce/control costs as much as possible but has the disadvantage of requiring the target cost to be fixed on the basis of a rough estimate. The variable fee concept is more illustrated in the following example:

Example 5.1

Bidding costs		Final completion costs	
Estimated cost	\$500,000.00	Final cost	\$510,000.00
Fixed fee	\$50,000.00	Fixed fee	\$50,000.00
Variable fee	10% of ± \$ 500,000.00 (10% on increases or decrease to estimated cost)	Applicable variable fee: - (\$10,000.00 × 10%)	–\$1000.00
Tender sum	\$550,000.00	Final account	\$559,000.00

Cost reimbursable contracts are suitable where Owner has trust in the Contractor and in situations where overall scope of work is not clear at initial stage and where increased frequency of additions and alterations is expected or on emergency work projects. Cost reimbursable contracts allow contractors early involvement at the design stage and allow Owner's participation in contract management.

5.3.4 *Guaranteed Maximum Cost Contract (GMC Contract)*

On some projects where the scope is well defined, the Owner and the Contractor agree to a project cost guaranteed by the Contractor as maximum, also known as a ceiling price. In this system Contractor takes all the risks, both in terms of actual project cost and project time. Thus, a guaranteed maximum cost arrangement imposes a penalty on a Contractor for cost overruns and failure to complete the project on time. With a guaranteed maximum price contract, any amounts below the maximum are typically shared between the Owner and the Contractor, while the Contractor is responsible for costs above the maximum.

Example 5.2
Consider an example in which the Owner agreed with the Contractor to the following arrangement:

Target cost: $ 500,000.00
Target profit: $ 50,000.00
Target total cost: $ 550,000.00
Sharing ratio agreed: 70/30 (owner 70%, contractor 30%)
Ceiling price: $ 600,000.00

Scenario-1 Suppose the cost of construction decreases or underran the target cost, the Owner and Contractor would split the savings according to the share ratio. More specifically, if the cost of work completed is $450,000.00 ($50,000.00 less than target cost), the Contractor's share is 30% of underrun, i.e., $15000.00. Hence, the Contractor will get:

Cost of work done ($450,000.00) + target profit ($50,000.00) + underrun share ($15,000.00) = 515,000.00.

Scenario-2 In this scenario, consider that the Contractor completes the work with cost overrun but below ceiling price, say at $520,000.00, overrunning the target cost by $20,000.00. The Contractor's share of overrun is 30% of $20,000.00 ($6000.00). The target profit will be reduced by this amount ($50,000.00–$6000.00) equal to $44,000.00. Hence, the Contractor will then get: $520,000.00 + $44,000.00 = $564 ,000.00, which is $36000.00 below ceiling price.

If the cost of work done exceeds and overran the ceiling price of $600.000.00, the Contractor will receive no profit and has to bear the additional costs.

GMC contracts provide the Owner a nice feeling of security. Entering into a contract of this nature, he is convinced that no matter what happens, the final cost will not be above the maximum and there is a fair chance it could be lower. Any design changes which result from the specific instructions of the Owner would understandably fall outside the guaranteed price.

GMC contracts best achieve the Owner's objective because a partnership is formed between the Owner and the Contractor wherein the Owner agrees to reimburse the Contractor for actual cost as it occurs, not from a schedule of rates. This eliminates the distrust between parties. It also eliminates some contractor's tendency to not pay his suppliers and subcontractors because he gets audited monthly. In today's market, this one issue alone will solve a lot of problems and insure both savings and a smoothly running project.

Benefits include that the Owner can play an active role throughout the entire process. The whole issue of cost is manageable when the savings are shared, rather than negotiated from an adversarial position. When the administration is properly setup and organized, the benefits are truly amazing. Because every purchase order and invoice received from the Contractor is submitted to the Owner as backup, and because the Owner agrees to cut the time for processing and pay promptly, a positive and successful relationship is encouraged. In general, GMC contracts are very similar to a lump sum contracts.

5.3.5 *Time-and-Material (T&M) Contract*

Under this contract, the Contractor is paid on the basis of actual cost of labor at fixed hourly rates, actual cost of materials and equipment used, and agreed-upon markup to cover the Contractor's overhead and profit. T&M is commonly known as "Force Account."

T&M contracts contain aspects of both contract categories (cost reimbursable and fixed price). They resemble fixed price type arrangements in that they are priced on fixed hourly rates, and they also resemble cost reimbursable type arrangement because they are open ended as the total cost of material and equipment is unknown. Work on a T&M basis is considered mostly during construction, when the scope of work cannot be well defined.

T&M contracts are not considered beneficial because the Contractor is paid for the number of hours actually used to perform the job and cost of material installed. Hence, the Contractor has no incentive to control material costs or manage the labor force efficiently. Therefore, proper surveillance on the Contractor is required to assure that the Contractor is performing efficiently and using effective cost control measures. Daily work records must be prepared either by the Contractor or by Contract Administrator reporting the labor and equipment employed and the material used and signed by both parties on a daily basis. It is also advisable that the contract shall include a ceiling or not-to-exceed price. Under this arrangement, the Contractor will be bound to charge for labor and material up to a certain maximum and will assume the excessive costs.

5.3.6 *Strengths and Weaknesses of Cost plus Contracts*

These contracts are faster, as scope or design need not to be defined completely. There is flexibility for changes during execution of the work. Use of guaranteed maximum cost types provide owners with some cost certainty. Additionally, owners have to pay only for the actual work performed and cost actually incurred.

The weaknesses include that the total cost is not well defined. It requires close inspection and review of construction. Since under cost plus contracts, the scope is not well defined, most of the risks are shifted to the Owner. Contractor also needs to develop adequate accounting systems. Certainty of cost is limited until the project is complete. Under T&M, the Contractor has no incentive to be efficient to control the cost; hence, more efforts for monitoring are required by the Owner.

References and Further Reading

1. CCDC-2. (2008). *Canadian Construction Document Committee- Stipulated Price Contract.*
2. CCDC-4. (2011). *Canadian Construction Document Committee – Unit Price Contract.*
3. The FIDIC. (1999). *Conditions of contract for construction – Red book.* Author: FIDIC.

Chapter 6
Selection of an Appropriate Contract Strategy

6.1 Introduction

Choosing an appropriate strategy for a particular project is a complex decision as it is a primary factor in determining the cost of and schedule for the project. This decision has become more difficult in recent years as several delivery methods have been developed in addition to the traditional method.

Contract strategy means selecting organizational and contractual policies, means and methods required for the execution of a specific project throughout all stages of pre-design, design, construction and post construction with a goal of meeting main project objectives.

Project contracting strategies depend primarily on the Owner's objectives and priorities which include a balance between a quality, schedule and cost. Also considered are contract method and pricing type, allocation of risks and responsibilities, finance, Owner's resources, etc.

Since we have previously discussed most of the above issues, we can set out the objectives and priorities that could be considered by the Owner for selecting an appropriate contract strategy for construction of any project. The following highlights the possible objectives and priorities that could be considered by the Owner in developing a contract strategy.

6.2 Project Scope

The project scope describes the objectives, specific tasks to be executed, costs and timeframe for their completion. It guides the project team about the extent of work to be carried out and forms the baseline against which potential changes are assessed

A. Surahyo, *Understanding Construction Contracts*,
DOI 10.1007/978-3-319-66685-3_6

and the project's performance is measured. The purpose of defining scope is to clearly describe the logical boundaries of the project. Scope statements are used to define what is within the boundaries of the project and what is outside those boundaries.

One of the primary causes for project failures is that the contracting strategy is selected without sufficient consideration for scope of the project and specific project conditions, which results in; conflicts between the contracting parties, scope creep, claims, disputes, and other undesirable results. It is therefore important that the scope of work should be well defined, comprehensive, and precise including all necessary details of the work. In describing the scope of work, efforts should be made to avoid generalization and vagueness.

Another cause of project failures is poor management of the project scope during construction. Since it is not possible to identify every requirement and feature that will be required for the final product; changing or refining scope while a project is underway is inevitable. It is wrong to resist all scope changes. Controlling and managing scope change is critical to the success of any project, as scope changes can significantly impact the cost, schedule, risks and quality of the entire effort.

One of the principal requirements for scope management is that all changes to the project scope must be approved by the Owner. Effective scope management requires good communication to ensure that everyone on the team understands the scope of the project and agrees upon exactly how the project goals will be met.

The traditional method with pricing options of lump sum and cost reimbursable method of guaranteed maximum cost offers the Owner a significant amount of control over the end product because the project features are fully determined and specified prior to selection of the Contractor.

The construction management method under CM-at-Risk also provides scope control to some extent as the CM takes the responsibility of quality work, its cost, and completion date.

6.3 Cost

Generally, it is accepted that an open tendering process will result in achieving the most competitive price from the Contractor. Some owners need to know the final contract value before they start a project. In this case, the design-build method with open tendering can give a firm, guaranteed, and competitive price.

The guaranteed maximum cost model is more suitable for cost control. For fast-track projects, price certainty is difficult to guarantee. Cost control can also be achieved by using traditional methods. It comes in a variety of forms, like measured quantities, approximate quantities (lump sum), or no quantities. The lump sum or fixed price contracts are the most suitable at controlling costs, but the final figure may be different due to changes or varying site conditions.

6.4 Time

The time required from inception to completion of a project will influence the choice of contract strategy. Many organizations and Owners require early/speedy occupation of the building or opening of a facility and, hence, wish to reduce the duration of a project. In such cases, fast-track methods like design-build or contract management method with two-stage tendering procedures allow a much earlier start on site and can best meet this objective.

Speedy construction can also be achieved with traditional methods by overlapping the various stages of designing and construction; however, this is generally not recommended.

6.5 Quality

Under the design-build method, there is no check and balance; hence, the quality achieved mainly depends upon the Contractor, particularly where the Owner does not employ any professional advisor. The contractor himself then regulates the quality of work.

Similarly within a fast-track approach to the contract management method, which utilizes a number of different specialist contractors on site, widely varying standards of workmanship can be encountered. The ability to control quality will vary since this will depend upon the type of subcontractors employed.

Design-bid-build or traditional contracting is the most suitable method to obtain controlled quality, and workmanship as design and construction are separately controlled. Usually, the construction works are supervised by the Designer's representatives. Additionally, the selective bidding holds the best solution to quality control as a limited number of suitable and reputed contractors are usually chosen to bid who take care to enhance their reputation by doing quality work.

6.6 Size of Project

Based on the level of complexity and uniqueness of the project, the owners need to maintain an appropriate level of control. The more complex and costly a project, the greater the need for professional management and advice.

Small size projects are not suited to the more elaborated forms of contractual arrangements, as they will not be cost effective. The Green Book of FIDIC Conditions of Contract [1] is a suitable form of contract for small size projects. Traditional contracting involves an organizational structure and is therefore too complex for small and simple projects. Small-sized projects, therefore, can use design-build (DB) contracting with either a selective or open tendering procedures.

With its single point of responsibilities, it would be better to rely on the expertise, skills, and judgment of a Contractor.

Medium-to-large or highly complex projects can use many contractual options. Traditional contracting can cope with large and complex projects but tend to be high in claims. For complex projects, construction management is the most suitable method of contract. This may be because Construction Managers tend to be more experienced in handling such projects and the Owner's full involvement and direct dealing results in a greater degree of control. DB is used successfully for both vertical (buildings) and horizontal (all modes of transportation) projects of varying size, cost, and complexity. The use of DB has been shown to save time in project duration, reduce project change orders, and increase the potential for innovative design. Complex problems are always solved more efficiently and with optimal solutions through collaboration. Collaboration occurs throughout a DB project, not only within the DB design/construction team but also with the Owner.

6.7 Owner Requirements

The requirement of the Owner greatly affects the choice of contract strategy. Some owners have in-house technical staff and prefer to be more involved in the process, whereas, others rely on a third party and retain a Consultant to manage the project.

The majority of projects awarded by the government and local authorities use traditional (design-bid-build) method with selective tendering. Some private organizations find management fee contracting more suitable to their needs.

Under the traditional method, most of the management part is delegated to the Consultant; hence, the Owner's involvement is reduced. However, the Consultant and the Contractor are separately and directly responsible to the Owner. In general, responsibility for the whole project is divided among all three parties involved.

Under the design-build method, only one party is responsible for the whole project, but the Designer is liable to the Contractor instead of Owner if any dispute arises. It is therefore suitable for an Owner to retain an independent project manager to look after his interests and who is directly responsible to him.

Under the Construction Management method, the role of the main Contractor is very limited and the Owner or Construction Manager remains fully involved in the project, resulting in a greater degree of control and responsibility.

6.8 Risk Allocation

The risk associated with a project is one of the primary factors influencing selection of a contract strategy. Risks are not always obvious, and may exist at all project levels throughout the life of a project. Starting any construction project involves the Owner taking risks because things may not go exactly as expected or planned.

Risk management issues are addressed in related Chapter- 9, whereas advantages and disadvantages of different contract types and methods have already been discussed in previous chapters. The following section will discuss the impact of risks and how they are distributed among various parties based on different types of contract.

A. CONTRACT METHODS

1. *Traditional methods*: With the traditional method, the risks are divided between the Owner and the Contractor. Under most of the standard forms, the Contractor takes the risk to complete the work in the stipulated time period and in accordance with contract terms. If he fails to meet these requirements, he will be liable to pay damages. He is also responsible for quality of works, materials, plants, and damage to persons until the project is taken over by the Owner. The responsibility for care of the work makes the Contractor responsible for the acts and defaults of every person on site, including any Subcontractor, his agents, or employees.

   The risks allocated to the Owner are usually a result of damage caused to an occupied portion of permanent works, loss or damage caused by design of works and forces of nature like earthquake, war, riots, floods, etc. The Owner is also at risk to the Contractor if he does not pay for the works done by the Contractor.

   The Consultant is at risk to the Owner to perform in accordance with the standard of professional care, and if he fails to meet these requirements, he will be liable to agreed damages. The Owner is also at risk to his Consultant, if he fails to pay for his services.
2. *Design and build methods*: Under the design-build method, the Contractor assumes the majority of risks. He is ultimately responsible for any defects or deficiencies in the work including any defects or deficiencies in design. This means, the risk of the cost exceeding the price lies entirely with the Contractor. The Contractor, being a single point of responsibility, is also liable for the performance, quality, safety, and completion of the project on time.

   The Owner will be liable for extra money only when he asks for changes or hinders the work progress. Since the Owner is not as involved in design and construction, he has less control over the finished project.
3. *Construction management methods*: Under *CM*-as-Agent, most of the risks of construction cost overruns and insurance risks remain with the Owner. Since the Owner contracts direct with the trade Contractors, the total risk in the event of failure or disputes lies with the Owner. However, if the *CM* fails in any of his duties, he will be liable to the Owner for the losses provided in the agreement between Owner and the *CM*. Under this path, the Owner also assumes the risk of becoming a "Constructor" under "Ontario Occupational Health and Safety Act" legislation [2]. He will be responsible to comply with the related provisions of all H&S legislation for all persons working on site.

Under the *CM*-at-Risk arrangement, the *CM* contracts with the Subcontractors/Trade Contractors and assumes the risk of completing the project within budget and stipulated time meeting required quality and health and safety standards.

B. CONTRACT TYPES

1. *Lump sum contract*: This contract type places maximum risk upon the Contractor and full responsibility for all costs and resulting profit or loss. It provides maximum incentive for the Contractor to control costs and perform effectively and imposes a minimum administrative burden upon the Owner. The primary risk factors to the Owner with this type of contract are:

   - Changes in scope resulting in payment adjustments on a noncompetitive basis.
   - Unforeseen complexities in field conditions that may result in change in quantities.
   - Differing site conditions.
   - Excusable delay conditions, i.e., delays that are contractually permissible, allowing the Contractor more time and possibly more money.
   - If quality expectations are not clearly defined, the contractor will likely take short cuts in order to complete the project as soon as possible.

2. *Unit price contract*: The primary risk factors to the Owner with this type of contract are:

   - Payment adjustments for quantity overruns.
   - Differing site conditions where the quantities are rather unpredictable.
   - Excusable delay conditions.

3. *Cost reimbursable contracts*: These contracts eliminate a large number of risks in the project for the Contractor and place them with the Owner. The Owner gains greater flexibility in a cost reimbursement contract, and the Contractor remains satisfied as he is liable to get payment for changed or unforeseen events.
4. *Guaranteed maximum cost contract*: The intention of this pricing route is to transfer all risks to the Contractor and allow for no increase in price whatsoever other than costs which result from Owner changes. A Contractor who undertakes a *GMP* contract has to take financial risks together with risks such as unforeseen ground conditions, unexpected encounters with underground utilities, bad weather, industrial unrest, shortage of labor, machinery and equipment, materials, changes in legislation, insolvency of suppliers and subcontractors, fire, storm, and earthquake. Contractors are even asked to check all information received from the Owner and take responsibility for its accuracy.

The risk factors to the Owner with this type of contract are that they require strong administration. The main effort involves defining what the cost is. The management representing the Owner should be efficient collecting discounts, obtaining credits for small tools, establishing realistic labor rates, and monitoring rentals and generally be really prepared to act as a partner with the Contractor in the project.

Based on the various identified risks associated with the method and types of contracts outlined, herein, the Owner can determine the most suitable contract delivery method for the project to meet his requirements.

6.9 Owner Expertise and Responsibilities

It is important for the Owner to have clear goals and objectives for his project, from the outset of the process. The Consultant and the Contractor must clearly understand the Owner's requirements in order to fulfill their obligations. Decision made by the Owner in selection of the Consultant and determination of the project delivery system are extremely important to the project and should be made carefully to obtain satisfactory results. It is important for the Owner that the design documents should be constructible, complete, and coordinated.

Decisions made during the preconstruction period have an impact throughout the project's duration. The Owner must thoroughly understand his preconstruction responsibilities and act accordingly. He should plan to expend appropriate time during the preconstruction phase of the project to enable the Consultant and Contractor to be provided with the necessary direction and information to meet their obligations.

While all parties involved in creating the completed project have definite responsibilities, the Owner bears the risk and has all the responsibilities for the work until he has contracted or assigned a specific responsibility to others. The Owner is responsible for providing the contractor Construction Documents that are adequate for the proposed construction.

When all above objectives and priorities are properly evaluated, a good decision can be made on the selection of an appropriate contract strategy for construction of any project.

References and Further Reading

1. FIDIC. (1999). *Short form of contract* (*Green book*). Published by International Federation of Consulting Engineers (FIDIC).
2. OHSA. (2014). Occupational health & safety act and regulations for construction projects.

Chapter 7
Construction Documents

7.1 Introduction

Construction document packages usually include two major groups of documents:

1. The bidding document package containing the necessary documentation for bidding the contract. The bidding process is further discussed in Chap. 8.
2. The Contract Document package containing the legal documentation to bind the contract between the Owner and the Contractor. The written document package is also known as the Project Manual.

Construction documents are basically the written and graphic documents, which communicate the design of the project, and govern administration of the contract for its construction. In general, the construction Contract Documents serve the following purposes:

(a) Defines the project delivery model and describe, in detail, what the project involves.
(b) Explains the scope of work and services to be performed.
(c) Identify materials to be incorporated into the project including their quality.
(d) Describes the contract obligations between the parties and the responsibilities of the Consultant who administers the construction contract for the Owner.
(e) Serves as a tool for controlling, monitoring, administering, and managing the project.
(f) Provide information for acquiring regulatory and financial approvals required for construction.

Complete and fully coordinated construction documents are a key component in effective performance, enhanced communication, and avoidance of disputes and litigation in today's world of design and construction. From a management and administrative point of view, every item within the Contract Document is important because it describes and records all aspects of the project.

A. Surahyo, *Understanding Construction Contracts*,
DOI 10.1007/978-3-319-66685-3_7

The Contract Documents normally comprise the following:

1. Contract forms
2. Conditions of contract
3. Specifications
4. Construction drawings
5. Bill of quantities/schedule of rates
6. Addenda

7.2 Contract Forms

Primarily, Contract Documents (particularly in Canada) take the following forms:

(i) Form of agreement
(ii) Performance bond
(iii) Labor and material payment bond
(iv) *WSIB* (Workplace Safety and Insurance Board) and tax statutory declaration
(v) Insurance certificate

7.2.1 Form of Agreement

This is the written document signed by the Owner and Contractor and comprises the legal binding between the parties to the contract. The agreement defines the relationships and obligations between the Owner and the Contractor. By reference, it incorporates the other Contract Documents including Contractor's submitted bid, addenda, and bid revisions.

Usually the agreement mentions that the Owner has accepted the bid by the Contractor for the execution and completion of such works in accordance with the terms and conditions of the Contract Documents. The Owner thereby covenants to pay the Contractor in consideration of the execution and completion of the works and remedying of defects therein, in accordance with the prices set out in the contract.

7.2.2 Performance Bond

A performance bond is an undertaking by a surety that the Contractor will perform and complete the contract. The performance bond is obtained to mitigate risk in the event the Contractor defaults or fails to perform. The surety will fulfill the Contractor's obligations in accordance with the terms of the contract in the event of default. This is further explained in Chap. 10.

7.2.3 Labor and Material Payment Bond

A payment bond guarantees the Owner that all bills for labor and materials contracted for and used by the Contractor will be paid by the surety to the vendors if the Contractor defaults. This also is further explained in Chap. 10.

7.2.4 WSIB *Declaration*

This declaration form is submitted by the contractor, certified by a government officer that the Contractor has paid all compensation owed to the "Workplace Safety and Insurance Boards" and also paid taxes provided for in the article of General Conditions of Contract. This declaration is submitted by the Contractor prior to the start of work. *WSIB* clearance is basically a number issued free of charge by *WSIB*. It declares that a business is registered with *WSIB* and has an account in good standing.

In Canada, The Workplace Safety and Insurance Board (formerly the Workers' Compensation Board) was established in 1997 by the Workplace Safety and Insurance Act, 1997, (WSIA). Under Part II of this Act, the Workplace Safety and Insurance Board (WSIB) has a mandate to promote health and safety in workplaces and to reduce the occurrence of workplace injuries and occupational diseases.

Under this Act, Employers contribute to a province-wide insurance fund. Contributions in the form of (insurance premiums) are based on the Employer's payroll and the accident frequency in their industry. Injured workers are compensated by the *WSIB* on a "no-fault" basis meaning no matter who is at fault, the Employer, the employee, or someone else. In return for automatic compensation, the Employer is shielded from any other liability. This means an employee cannot sue his Employer for negligence if that negligence causes a work-related injury or disease. However, it is the *WSIB* who decides, according to the Act, whether or not a worker receives compensation, and decides what the worker gets compensation for.

Most WSIB claims are straightforward. The injured worker is paid benefits while recovering and returns to his or her regular work. But some cases are complicated and may take time to resolve. Some injured workers are denied with *WSIB* benefits or have their benefits stopped by the WSIB because the Employer has the right to contest an employee's *WSIB* claim. For example, an employee might have a back injury which he says was caused on the job. The Employer may claim that the injury was caused in a car accident or somewhere else.

The main concept of the *WSIB* system is to get the injured worker back to his or her work as soon as practicable. The worker and the Employer must cooperate toward this goal. If the injured worker cannot perform his or her regular duties, the Employer may offer suitable modified work that the worker can perform without aggravating the injury. If the work is suitable, then the injured worker must accept it.

7.2.5 *Insurance Certificate*

Contractors are required to provide evidence of all insurance coverage as part of the terms and conditions of contract prior to start work. This certificate is issued by an insurer who certifies the existence of insurance coverage under specific conditions. More specifically, the document lists the validity date of the policy, the type of insurance coverage purchased, and the types and dollar amount of applicable liability. Construction insurance details are further discussed in Chap. 10.

7.3 Conditions of Contract

In the construction industry, the conditions of contract define the basic rights, responsibilities, and relationship of the parties involved or the rules by which each party must comply. Conditions of contract mostly consist of:

7.3.1 *General Conditions*

General conditions contain general clauses that establish how the project is administered and are intended to be used unchanged for every project. It is usually in the form of published standard document that include written principals common to most construction contracts.

7.3.2 *Particular/Supplementary Conditions*

Supplementary conditions are specially prepared to modify or supplement the general conditions as needed to accommodate the unique requirements of a specific project.

The conditions of contract form a very important part of the Contract Documents; hence careful attention must be paid, to the written agreements that Consultants, Owners, and Contractors enter into. Standard general conditions of contract are mostly prepared to the mutual benefit of all parties to the contract as they establish measured and predictable standards among the parties. Standard forms are based on generally accepted professional and industry norms and are fair and balanced.

Another reason for using standard general conditions is that they provide language and provisions that have been tested in practice as well as in the courts. They are well understood and familiar to all parties involved in the construction. The standard forms are amended periodically as legislation and conditions change, thereby making improvements for both the Owner and the Contractor. The standard forms are usually drafted by committees representing professional experts in construction industry, lawyers, and financial/insurance experts.

There are, in practice, several private and public forms of general conditions of contract that have been established by various large organizations for their own use. The impositions of conditions of contract which are biased in favor of the Employer are not recommended.

7.4 Standard Forms of Conditions of Contract

There is variety of standard forms of contract in use in the construction industry. The selection of a particular form will depend upon the circumstances surrounding the project and size of the project. Some of the standard general conditions for construction prepared and published by professional organizations are:

7.4.1 Canadian Construction Documents Committee (CCDC)

The CCDC is comprised of Owners (two members from the private sector and two members from the public sector), the Association of Consulting Engineering Companies – Canada (three members), the Canadian Construction Association (four members), Construction Specifications Canada (three members), the Royal Architectural Institute of Canada (three members), and an ex officio member from the legal profession. The various documents published by CCDC are:

7.4.1.1 CCDC 2 – 2008 Stipulated Price Contract

This general conditions document can be used between Owner and Prime Contractor to govern the required work for a single, predetermined fixed price or lump sum contract.

7.4.1.2 CCDC 3 – 1998 Cost Plus Contract

This general conditions document can be used between Owner and Prime Contractor to govern the required work on an actual-cost basis, plus a percentage or fixed fee which is applied to actual costs.

7.4.1.3 CCDC 4 – 2011 Unit Price Contract

This general conditions document can be used between Owner and Prime Contractor to govern the required work for a predetermined, fixed amount for each specified unit of work performed. The total price is determined by multiplying the unit price by the actual, measured quantity of work performed for each specified unit.

7.4.1.4 CCDC 5A – 2010 Construction Management Contract: For Services

This general conditions document can be used between Owner and Construction Manager for which the work is to be performed by Trade Contractors. The Construction Manager acts as a limited agent of the Owner providing advisory services and administering and overseeing the contracts between the Owner and Trade Contractors.

7.4.1.5 CCDC 5B – 2010 Construction Management Contract: For Services and Construction (CM-at-Risk)

This general conditions document can be used between Owner and Construction Manager to provide advisory services during the preconstruction phase and perform the required work during the construction phase. At the outset, the Work is performed on an actual-cost basis, plus a percentage or fixed fee which is applied to actual costs. The parties may agree to exercise the following options: guaranteed maximum price (GMP), *GMP* plus percentage cost savings, or conversion into a stipulated price contract.

7.4.1.6 CCDC 14 – 2013 Design-Build Stipulated Price Contract

This general conditions document can be used between an Owner and Design-Builder where the Design-Builder performs Design Services and Construction under one agreement, for a single, predetermined stipulated or fixed price.

7.4.1.7 CCDC 17 – 2010 Stipulated Price Contract for Trade Contractors on Construction Management Projects

This general conditions document can be used between an Owner and Trade Contractor when performing work for a single, predetermined fixed price, regardless of the trade Contractor's actual costs. It is specifically for use where the project is performed under the CCDC 5A Construction Management method of contracting.

7.4.1.8 CCDC 18 – 2001 Civil Works Contract

This general conditions document can be used between Owner and Contractor for civil works construction, e.g., roads, bridges, dams, underground utilities, etc.

CCDC also produces guides to assist with using and understanding the aforementioned contract types. Other professional organizations in Canada have similar publications available to assist the user, such as:

- Association of Consulting Engineering Companies (ACEC)
- Canadian Construction Association (CCA)
- Construction Specifications Canada (CSA)
- Royal Architectural Institute of Canada (RAIC)

7.4.2 FIDIC: International Federation of Consulting Engineers

Founded in 1913, the FIDIC (International Federation of Consulting Engineers) with headquarters in Geneva, Switzerland, is charged with promoting and implementing the consulting engineering industry's strategic goals on behalf of its Member Associations and to disseminate information and resources of interest to its members. Today, the FIDIC membership currently covers 100 countries of the world.

The FIDIC, in the furtherance of its goals, publishes international standard forms of contracts for Owners, Consultants, Sub-consultants, joint ventures, and representatives, together with related materials such as standard prequalification forms. In 1999 the FIDIC updated its standard forms of conditions of contract, with the publication of a first edition of a new family of contracts comprising major four new standard forms of contract, usually known by their color:

(i) *FIDIC Conditions of Contract for Construction* (Red Book)
(ii) *FIDIC Conditions of Contract for Plant and Design-Build* (Yellow Book)
(iii) *FIDIC Conditions of Contract for EPC/Turnkey Projects* (Silver Book)
(iv) *FIDIC Short Form of Contract* (Green Book)

7.4.2.1 Red Book

This is the most widely used form of contract throughout the world. The Red Book is drafted for the use of building and engineering works. The Red Book contains three parts:

Part 1: General conditions of contract. This part comprises Clauses 1–20 together with the Appendix and Annex for dispute adjudication board agreements.
Part II: Guidance for the preparation of particular conditions.
Part III: Forms – this part includes examples of the letters and agreements which are referred to in the general conditions.

7.4.2.2 Yellow Book

The FIDIC recommends use of the Yellow Book when building an electrical and/or mechanical plant and building and engineering works designed by the Contractor. Under the usual arrangements for this type of contract, the Contractor designs and provides, in accordance with the Employer's requirements, plant and/or other works, which may include any combination of civil, mechanical, electrical, and/or construction works.

7.4.2.3 Silver Book

Silver form is applicable for international major turnkey projects, with the idea that responsibility for all work (engineering, procurement, and construction (EPC)) is to be taken by the Contractor who completes the finished product ready for operation (at the "turn of the key").

7.4.2.4 Green Book

The Green Book is recommended for engineering and building works of relatively small capital value. However, depending on the type of work and the circumstances, the Green Book may be suitable for contracts of considerably greater value. They are considered most likely to be suitable for simple or repetitive work or work of short duration without the need for specialist subcontracts. This form may also be suitable for contracts which include, or wholly comprise, contractor-designed civil engineering, building, mechanical, and/or electrical works.

7.4.3 NEC (New Engineering Contract) Contracts, UK

NEC is produced by the Institute of Civil Engineers UK (ICE). NEC published its third edition, NEC3, in 2005 and introduced more recent editions in 2013. NEC3 is endorsed by many organizations nationally and internationally.

NEC3 provides a wide range of Contract Documents and can be used for civil, building, nuclear, utilities, infrastructure, facilities management, oil and gas, purchasing, and supply. NEC3 main versions of Engineering and Construction Contracts (ECC) are:

(i) Option A: Priced contract with activity schedule
(ii) Option B: Priced contract with bill of quantities
(iii) Option C: Target contract with activity schedule
(iv) Option D: Target contract with bill of quantities

(v) Option E: Cost reimbursable contract
(vi) Option F: Management contract

7.4.4 *Infrastructure Conditions of Contract, UK*

Infrastructure Conditions of Contract has replaced original *ICE* (Institute of Civil Engineering, UK) *Conditions of Contract* on 1 August 2011. The ICE has transferred its ownership of ICE Conditions of Contract to the ACE and CECA. ACE (Association of Consulting Engineers) and CECA (Civil Engineering Contractors Association) now solely own the *Infrastructure Conditions of Contract.*

The *Infrastructure Conditions of Contract*, a standard suite of forms of contract, is mainly based on the highly successful ICE Conditions of Contract and comprises following main versions:

(i) Measurement Version – 01 August 2011
(ii) Design and Construct Version – 01August 2011
(iii) Target Cost Version – 01 August 2011
(iv) Term Version – 01August 2011
(v) Minor Works Version – 01August 2011

7.4.5 *JCT (Joints Contracts Tribunal) Contracts, UK*

This form of contract is produced by the following seven members (UK):

(a) Royal Institute of British Architects (RIBA)
(b) Royal Institute of Chartered Surveyors (RICS)
(c) British Property Federation (BFP)
(d) Contractors Legal Grp. Limited
(e) Local Government Association
(f) National Specialist Contractors Council
(g) Scottish Building Contract Committee

The JCT suite of contracts' most recent version was published in 2011. The main JCT contract forms are:

(i) Standard Building Contract
(ii) Design and Build Contract
(iii) Construction Management Contract
(iv) Management Building Contract
(v) Minor Works Building Contract
(vi) Intermediate Building Contract
(vii) Major Project Construction Contract (MP11)

7.4.6 *Engineers Joint Contract Documents Committee (EJCDC) Documents, USA*

These documents are prepared by Engineers Joint Contract Documents Committee (EJCDC) which includes:

(a) American Council of Engineering Companies
(b) American Society of Civil Engineers
(c) National Society of Professional Engineers
(d) Associated General Contractors of America and more than 15 other organizations

The family of documents published by EJCDC is based on various series as follows:

(i) Construction family of document (C-Series)
(ii) Engineering family of documents (E-Series)
(iii) Environmental remediation family of documents (R-Series)
(iv) Procurement family of documents (P-Series)
(v) Design and build family of documents (D-Series)

The EJCDC has released new editions of the Construction Series (C-Series) documents recently in 2013. The documents in the new C-Series are specifically written for use on public and private engineered facility projects. They are most commonly used for infrastructure and public works construction in the United States.

The most commonly used document is C-700, Standard General Conditions of the Construction Contract (2013), which is based on the traditional contract method, i.e., design-bid-build.

7.4.7 *American Institute of Architect (AIA) Documents, USA*

There are nearly 200 forms and contracts comprising AIA Contract Documents. These forms and contracts define the relationships and terms involved in design and construction projects. Prepared by the AIA with the consensus of Owners, Contractors, attorneys, Architects, Engineers, and others, these documents have been finely tuned during their 120-year history. As a result, AIA comprehensive contracts and forms are now widely recognized as the industry standard. AIA organizes Contract Documents by two methods:

By "families" based on types of projects or particular project delivery methods

By "series" based on the use of the document

By Families Based on families, AIA Contract Documents are divided into nine families based on project type or delivery method. Documents in each family provide a consistent structure and text base to support the major relationships on a design and construction project. Understanding AIA document families helps in selecting the most appropriate standard forms for the project.

- *Conventional (A201) family*: Conventional (A201) is the most commonly used family of AIA Contract Documents because the documents are suitable for the conventional delivery approach of design-bid-build. These documents can be used on small to large projects. The keystone document under this family is A201–2007, General Conditions of the Contract for Construction.
- *Construction Manager as Adviser (CMa) family*: The main document under this family is A232–2009 (formerly A201CMa–1992), General Conditions of the Contract for Construction, Construction Manager as Adviser Edition.
- *Construction Manager as Constructor (CMc) family*: The main documents under this family are coordinated for use with AIA Documents A201–2007, General Conditions of the Contract for Construction, and B133–2014, Standard Form of Agreement between Owner and Architect.
- *Design-Build family*: The main document under this family is A141–2014, Standard Form of Agreement between Owner and Design-Builder.
- Based on family setup, other AIA Documents are Integrated Project Delivery (IPD) family, Interiors family; International family, Program Management family, and Small Projects family.

By Series Based on series, AIA Contract Documents are classified alphabetically by document use or purpose.

- A-Series: Owner/Contractor Agreements
- B-Series: Owner/Architect Agreements
- C-Series: Other Agreements
- D-Series: Miscellaneous Documents
- E-Series: Exhibits
- F-Series: (Reserved)
- G-Series: Contract Administration and Project Management Forms

7.5 Specifications

Specifications define the qualitative requirements for products, materials, and workmanship for a proposed project. Many public authorities and consulting firms publish standard format of "general specifications," which establish a uniformity of standard work to be produced. Specifications usually describe the type and quality of products required for the project.

Specifications usually contain following information for prospective bidders:

(a) General requirements specifying contractual procedures and performance of work by a Contractor
(b) Technical specifications covering construction of the particular work to be performed

Buildings, highways, bridges, sewerage, and water works are some examples of the types of improvements for which agencies may have standard specifications. In general, the quality of materials and the standards of workmanship to be provided by the Contractor must be clearly described. Details must be included of samples to be provided and tests to be carried out by the Contractor during the course of the contract.

Generally, there are two basic types of specification: (1) performance based and (2) prescriptive based [1].

7.5.1 Performance Based

Under this method, the required end results are specified along with criteria by which the performance of a product, system, or material will be judged and the method by which it can be verified. Performance-based specifications focus on outcomes or results rather than process and the required goods and services rather than how the goods and services are produced. The Contractor is free to choose the materials and methods that comply with the performance specification. Performance specifications are designed to reduce Owner risk by transferring responsibility to the Contractor.

Example Paint: paint all damaged equipment with products matching original finish and quality in appearance.

7.5.2 Prescriptive Based

Prescriptive specifications are the most common method of specifying construction work results. Prescriptive specifications convey the requirements of a project through a detailed explanation of the materials that must be used and the activities and procedures of installing those materials to achieve the end result. The prepared specifications are mostly a combination of at least two and sometimes all three of the specification styles mentioned below. Most specifications also have a combination of prescriptive and performance specifications. Prescriptive specifications are further subdivided in main three specification styles:

- *Descriptive specifications*: Under this method, the required properties of the materials and methods of installation are described in detail without using proprietary or manufacturer's names. A descriptive specification is usually used to describe properties of complicated components or systems that cannot be adequately shown on drawings. Many industry standards, produced by trade associations or by standards association writing organizations, may be descriptive specification. However, these standards are normally included in a Project Manual by reference only.

  Example: epoxy Grout – premixed, nonshrink, consisting of thermosetting resin base, with inert fillers, with minimum 7-day compressive strength of 10,000 psi, suitable for use on dry or damp surfaces.

- *Proprietary specifications*: A proprietary specification is one which requires the use of a sole source product. A specification, under this method, describes a product, material, assembly, or piece of equipment by trade name and manufacturer name that produces products acceptable to the owner or design professional.

  Example: Interlocking pavers – shall be 80-mm-thick "Uni Eco-Stone or Ecoloc" as manufactured by Unilock Ltd.

  The basic reason for using proprietary specifications is to control product selection. When a product is required for which no consensus standards exist, or for which special qualities are desired, a proprietary specification can be used. The main disadvantage of using proprietary specifications is they eliminate competition resulting in high prices.

- *Consensus standards*: These standards are developed by an authority, custom, or general consent with a recognized and accepted criteria. Under this method, the technical characteristics of certain aspects of the construction are defined by reference to standards. The specified standards should be internationally accepted and widely used. The standards to be applied should be clearly identified in the contract specifications. Usually standards define minimum criteria.

  Example: Portland cement – conforming to CAN/CSA- A3001 normal, Type GU Portland Cement.

  Standards used in the Canadian construction industry are mostly published by Canadian standards writing organizations, such as Canadian Standard Association (CSA), Canadian General Standard Board (CGSB), Canadian Gas Association (CGA), etc. These standards are generally endorsed by Standard Council of Canada (SCC). SCC is a federal government agency established by an Act in parliament to foster and promote voluntary standardization in Canada.

  The Canadian construction industry also uses standards that are written by US-based standards writing associations, such as The American Society for Testing and Materials (ASTM). The SCC equivalent standards endorsement organization in the United States is American National Standards Institute (ANSI). Both SCC and ANSI are authorized by the North American Free Trade Agreement to exchange and share standards freely and to share their respective authorized testing and certifying organizations. Both SCC and ANSI are ISO (International Organization for Standardization) members [1].

The specifications for any activity should describe the material requirements as well as the method of whole process. For example, the "Standard Specifications for Structural Concrete" prepared by American Concrete Institute (ACI) committee 301-96, outlines materials and proportioning of concrete, reinforcing and prestressing steels, production, placing and curing of concrete, and formwork design and construction. It also describes the methods of treatment of joints and embedded items, repair of surface defects, and finishing of formed surfaces.

Some of the renowned institutions who produce standard specifications are Construction Specifications Canada (CSC), Construction Specifications Institute USA (CSI), Canadian Standard Association (CSA), British Standard Institute (BSI),

American Concrete Institute (ACI), American Society for Testing Materials (ASTM), American Association of State Highway and Transportation Officials (AASHTO) etc., which most of the organizations incorporate in any required project applications. Standard specifications, published periodically, may be updated in the interim by issuance of amendments, revisions, or supplements.

CSC and CSI have jointly developed a consensus document called "MasterFormat" [2]. It is the most widely used standard for organizing specifications for building projects in the United States and Canada. Mainly, there are two documents which are being used in the construction industry for organizing construction information: MasterFormat [2] and UniFormat [3]. UniFormat is also jointly produced by CSC and the CSI.

MasterFormat MasterFormat [2] organizes information by work results (material and methods) with the primary purposes of organizing specifications with a standardized system of numbers and titles. The number and titles are grouped under two general headings: (1) Procurement and Contracting Requirements Group (Division 00) and (2) Specifications Group (Divisions 01–49) such as Division 01: General Requirements; Division 02: Site Construction; Division 03: Concrete; Division 04: Masonry; and so on. Specification sections are numbered using six digits or more; the first two digits are represented by division number.

It may be noted that MasterFormat does not include pre-written specifications. For specification templates that can be used to prepare project specifications, Spex.ca or CSI may be contacted.

UniFormat [3] This format is used to provide construction information based on physical parts or elements of a facility. UniFormat is not divided into divisions but is divided into eight major elements as follows:

A – Substructure
B – Shell
C – Interiors
D – Services
E – Equipment and furnishings
F – Special construction and demolition
G – Site work
Z – General

7.6 Drawings

The drawings basically convey the intentions of the Designers, Engineers, Architects, or Draftsman to the Contractor. The drawings are graphic representation of the work upon which the contract is based. They show the materials and their relationship to one another, including sizes, shape, locations, and connections.

The drawings must be in sufficient detail to enable Contractors to assess accurately, in conjunction with the specification and bill of quantities, the nature and scope of work included in the contract. It must be ensured that on all drawings for any one contract, the same conventions are adopted and uniformity of appearance and size is achieved, making the drawings easier to read. The element detailing must comply with any of the standard code requirements. For example, for dimensioning sequencing, length should always be given first, width second, and thickness third. Additionally, the dimensioning sequence and scale employed should be consistent throughout.

Construction involves many forms of drawings, such as:

(a) *Bid or tender drawings*: Tender drawings together with the other bid documents describe the project scheme to the Contractor so that he can price the construction work accordingly.

(b) *Construction or working drawings*: Construction drawings are usually same as tender drawings; however some tender drawings may carry amendments based on approved addendum. Construction drawings used on site therefore should carry stamp "issued for construction" to avoid confusion between tender drawings and construction drawings.

    Sometimes additional drawings are issued during construction due to changes. The Contractor and Consultant site staff must ensure that they have an adequate filling system for all of their drawings, which is of particular importance when recording changes. Superseded drawing copies should be clearly marked. They should not be destroyed or discarded until the final account has been agreed, since they may contain relevant information used for contractual claims.

(c) *Shop drawings*: During construction, the Contractor is required to submit shop drawings for the items specified in the Contract Documents. Shop drawing means drawings, diagrams, illustrations, schedules, performance charts, brochures, and/or other data which illustrate details of a portion of work. Shop drawing must indicate references to design drawings and specifications. These shop drawings are then reviewed by the consultant and returned as "R" (reviewed) or "RN"(reviewed as noted) or "RR"(revise and resubmit) as the case may be.

(d) *As-built drawings*: As-built or record drawings are the final set of drawings produced at the completion of a construction project. The construction drawings are continuously updated during the course of construction to mark any field changes of dimension and detail or changes to utilities etc. They include all the changes that have been made to the original construction drawings, including notes, modifications, and any other information that have been specified in Contract Documents or Contractor intends to include.

7.7 Bill of Quantities (B of Q)

The B of Q or schedule of rates comprises a list of items of work to be carried out, providing a brief description and estimated quantities of the work under contract. This is usually required under unit rate contracts. It allows each bidder to price the same information. The bill may include firm or approximate quantities depending upon the completeness of the drawings and other information based on how it was prepared. Calculation of the quantity of each item is prepared from the drawings and specifications.

Along with bidding facilities, the B of Q is also useful for other purposes, such as:

(a) Preparation of interim valuations for work under construction
(b) Ordering of materials, based on available estimated quantities
(c) Preparation of final accounts
(d) To assist in obtaining quotations from Subcontractors for sections of the measured works

Contracts based on stipulated prices usually have a schedule of prices including contingency to cover alterations, extras, and deductions. These schedule of prices are very helpful in determining pricing for changed items. Usually very few items are added/provided in schedule of prices which are not enough to properly address valuation for changes made during construction. Hence, calculating pricing for changes can create conflict between the Contractor and Consultant. It is therefore advisable to add pricing for as many items as possible within schedule of prices based on the work required. Since it will be priced by the Contractor during bidding, valuation for most of the changes can be agreed by both the parties.

7.8 Addendum

An addendum is the revisions of the bid documents made after the call for bids but before the bid closing. The main purpose of an addendum is to add, delete, or amend bidding documents, such as revisions to bid closing date, changes in the contract drawings and specifications due to errors or omissions, clarifying questions raised by bidders, etc.

Addenda (plural for addendum) should be delivered sufficiently in advance of the bid-opening date to permit all the proposed bidders to make the necessary adjustments in their bids. Addenda should be issued to all parties who purchased the bid documents. Bidders must acknowledge receipt of all addenda; otherwise their bids may not be accepted.

Faxing or emailing addenda shall be permissible, as long as record is kept, verifying that fax or emails were delivered to the recipients [1]. Addenda will become part of the Contract Documents when the construction contract is executed.

References and Further Reading

1. Notes from a course. (2014). *Principles of construction documents.* Attended by the Author – SAIT Polytechnic School of Construction, Calgary Alberta.
2. MasterFormat. (2011). *Master list of numbers and titles for the construction industry.* Jointly Published by Construction Specifications Canada (CSC) and The Constructions Specification Institute USA (CSI).
3. UniFormat. (2010). *A uniform classification of construction systems and assemblies.* Jointly Published by Construction Specifications Canada (CSC) and The Constructions Specification Institute USA (CSI).

Chapter 8
Bidding Procedures

8.1 Introduction

The basic structure of the bidding process consists of the formulation of detailed plans and specifications of a project based on the objectives and requirements of the Owner, and the invitation of qualified or negotiated Contractors to bid for the right to execute the project.

The aim for inviting bids is to obtain a competitive offer for the required project and to select a suitable Contractor. The offer becomes the base of entering into a contract with the successful bidder. The bid is the most important document submitted by the bidder, in which each bidder confirms that he has read and understood the requirements of the bid documents, and, based on such requirements, he enters his bid sum for undertaking and fulfilling all obligations under the contract.

The selection of contractors to bid is usually controlled by prequalification procedures or by developing a list of suitably qualified contractors for various sizes of contracts and types of work. For major projects and international contracts, prequalification of bidders is desirable since it enables the Owner or Consultant to establish, in advance, the competence of firms invited to bid. The prequalification gives added encouragement to contractors to respond to invitations to bid for those projects which they are best qualified to undertake.

The aim of prequalification process is to draw up a list of selected bidders while ensuring that proper competition is maintained. A further advantage of prequalification, which is beneficial to contractors and owners, is that unnecessary work involved in the preparation and subsequent evaluation of bids from inappropriate companies be avoided. For complete procedure of prequalification, "Standard prequalification form for contractors" published by the FIDIC (International Federation of Consulting Engineers) and "Contractors qualification statement – CCDC 11" published by the Canadian Construction Document Committee can be referenced.

A. Surahyo, *Understanding Construction Contracts*,
DOI 10.1007/978-3-319-66685-3_8

8.2 Contract "A" and Contract "B"

According to Canadian law (Judge-made law) of competitive bidding, the bidding process creates two types of contracts: Contract "A," the bidding contract, and Contract "B," the actual construction contract.

Contract "A," a binding bid contract, is created between each bidder and Owner upon the submission of a compliant bid. Contract "B" is the supply of material and services contract itself and is created when one of the bid is accepted by the Owner.

Generally the following fundamental principles are established in the law of competitive bidding:

1. If the Owner fails to comply with any terms and conditions set out in the bid, e.g., being unfair, the bidder can claim that the Contract "A" is breached.
2. If the bid documents call for irrevocable bids, a bidder has no right to withdraw or change its bid during the specified acceptance period [1]. The bid may be withdrawn prior to the bid closing date; however, after the closing date, each submitted bid shall be irrevocable and binding for a specified period (most of the Contract Documents recommend 90–120 days; CCDC 23 [1] recommends 30 days).
3. Bid submissions must be compliant with the bid documents, as privilege clause does not allow the Owner to accept non-compliant low bid as explained in Chap. 1.
4. When a bid is accepted by the Owner, the bidder must enter into a construction contract, i.e., Contract "B." The legal obligations of the successful bidder will be to perform the contract in accordance with the Contract Documents.
5. Owner must act in good faith and should treat all bidders fairly and evaluate the bids in a fair way. He should not accept non-compliant bids and shall not award contract on undisclosed criteria.

8.3 Bid Documents

The bid documents issued to the bidders are typically comprised of the following:

1. *Bidding Requirements*
   - Invitation to bid
   - Instructions to bidders
   - Information available to bidders
   - Bid security/bond
   - Bid forms
   - Supplements to bid forms
2. *Contract Documents*
   - Contract forms

- Conditions of the contract
- Specifications
- Drawings
- Bill of quantities
- Addenda
- Contract modifications

8.4 Bidding Requirements

Bidding requirements help bidders follow established procedures and submit bids that will not be disqualified because of technicalities. Bidding requirements address all prospective bidders interested in the project, while only the Contract Documents become part of the contract between the successful bidder and the Owner. Contract Documents were discussed in the previous chapter; here bidding requirements and procedures will be explained.

8.4.1 Invitation to Bid/Bid Solicitation

Invitation to bidders must be advertised in the largest circulation national newspaper. If international advertising is required, the advertisement should be placed in construction-related magazines or other periodicals, so as to attract the maximum reasonable number of suitably qualified bidders. A letter of invitation to bid typically carries the following information:

(a) Name of project and Owner with address
(b) Description of the works, goods, or services
(c) Type of bid (e.g., lump sum or unit price)
(d) Date, time, name of person, and place where tender is to be collected and submitted
(e) Date, time, location, and procedure of tender opening
(f) Price of documents charged, if any
(g) Bid security/bond value and requirements
(h) Owner's right to reject bids
(i) Date and time for pre-bid meeting or site visit requirements

The invitation often states that the Owner is not bound to accept the lowest or any tender or to be responsible for any costs incurred. Besides having clear documentation, contractors should be allowed an appropriate amount of time to schedule estimating work, review documents, and clarify questions in order to submit a responsible bid. It is generally acknowledged that a fair and responsive bid is better for both the Owner and Contractor rather than recovering from bidding errors.

The bid document issue form should be used to record which contractors have picked up the documents and at what time. The bid document issue form should be closed at the deadline for receipt of bidders and signed by the officer in charge of the bidding process.

8.4.2 Instructions to Bidders

The documents provided to the prospective bidders typically include instructions to bidders conveying information about the preparation, contents, submission, and evaluation of tenders, as well as model forms of the documents, which are to be submitted by the bidder with the bid. The instructions should specify all of the Owner's requirements in respect of the bids, including the criteria they must meet to be successful.

Instructions to bidders must be prepared by the Owner and Consultant to meet the requirements of individual contracts. The instructions to bidders should include requirements about alternative bids – the instructions to tenderers should state whether alternative bids will be considered or not. Alternative bids may suggest alternative designs, alternative construction programs or methods, and sequence of work, which may result in shortening the contract period. Alternative bid must be submitted with complete details of conditions of contract, drawings, specifications, calculations, and price of the portion of works that are to be altered. In practice an alternative bid would be considered only if it is found technically, contractually, and financially acceptable.

The instructions to bidders should also include information about the rules for rejection of bids and the procedures dealing with bid queries.

8.4.3 Information Available to Bidders

Information regarding toxic and hazardous materials, subsurface conditions, underground utilities, topographical and geotechnical surveys, etc. which may affect bidder's price should be made available by the Owner and should be issued with the bid documents to assist bidders in preparing their bids and to avoid disputes during construction.

8.4.3.1 Site Visits

Site visits are a prudent means for the prospective bidders to get a clear picture of the site before bidding. The arrangements for a site visit should be determined so that equal opportunities are open to all bidders. Site visitors should obtain information about variety of conditions, the precise nature of which depends on the work involved. Some examples are as under:

- Location of the project
- Site access
- Availability of electricity, water, telephone, and other services
- Surface topography, drainage, and so on

The Owner should record the firm/name of bidders who visited the site. Owner may require bidders or their representatives visiting site to indemnify the Owner/Engineer against any potential claims for damage, injury, or death as a result of site visit.

8.4.3.2 Currencies and Payments

Usually bidders are required to submit their bids in a single currency of that country in which the works are to be carried out. However, this information must be clearly mentioned in the bid documents. If more than one currency for payment is to be used, all bidders must be notified. In the case of one or more currencies, the rates of exchange to convert the local currency should be defined.

8.4.3.3 Bid Modifications

The bidders should be informed about the procedures for bid modifications. Usually, submitted bids may be amended prior to the specified closing date. Each bidder should examine the bidding documents carefully and not later than at least 50 days prior to the bid closing date and should make written request to the Consultant for interpretation or correction of any ambiguity, inconsistency, or error therein which he may discover. Any interpretation or correction shall be issued as addendum by the Consultant at least 7 days in advance of the closing date (timelines may vary depending on the Contract Documents). The issued addenda becomes part of the bid documents.

If the element of time is restrictive, the Owner may extend the deadline for submission of bids. Any addenda or extended deadline so issued forms part of the bid documents and is binding upon bidders. It is the responsibility of the Owner to make sure that all bidders should receive the amendments carried out.

8.4.3.4 Arithmetic Errors in Pricing

Bidders should be clearly informed about the procedure for dealing with arithmetic errors in pricing of bids. Frequently, the following procedures are adopted in dealing with such errors:

(a) If there is a discrepancy between the unit price and the total price that is obtained by multiplying the unit price and quantity, the unit price shall prevail and the total price shall be corrected, unless in the opinion of the Owner there is an obvious misplacement of the decimal point in the unit price, in which case the total price as quoted shall govern and the unit price shall be corrected.
(b) If there is an error in a total corresponding to the addition or subtraction of subtotals, the subtotals shall prevail and the total shall be corrected.
(c) If there is a discrepancy between words and figures, the amount in words shall prevail, unless the amount expressed in words is related to an arithmetic error, in which case the amount in figures shall prevail subject to (a) and (b) above.

8.4.3.5 Late Bids

A note should state that any bid presented to the Owner/Consultant after the prescribed deadline date and time for the submission of bids will be treated as non-compliant and will not be accepted. Bids received after the prescribed bid closing time should be returned to the bidder unopened, with "late bid and unopened" noted on envelop.

8.4.3.6 Pre-bid Queries

Interested bidders are allowed to request for clarification after the invitation for bid. The purpose of pre-bid queries is to clarify any concerns bidders may have with the bidding documents, scope of work, and other details of the requirement. The bidders should be informed in detail the format (e.g., letter, email, fax, etc.) in which queries will be accepted by the Owner. Sometimes Pre-bid meeting is held to clarify bidders concerns. These meetings are formal, and results are made available in writing to all prospective bidders. The time limit for the pre-bid queries should also be stated in the bid documents (FIDIC [2] recommends 30 days before the tender closing date).

Usually there are two methods for answering queries – correspondence method and conference method. In correspondence method, the bidders have to submit their query in the approved format to the Owner or consultant. The Owner or Consultant will reply to the query in writing, ensuring that it is also issued to all other bidders. Usually the identity of the bidder who had made the query is kept confidential.

In conference method, the Owner or Consultant will prepare a reply to all queries raised by various bidders and arrange a conference/meeting of the prospective bidders, where it should be read out by the Owner or Consultant, retaining the anonymity of the questioners.

Where the answer to a query is likely to result in additional work in compiling bids or where the answer to a query is delayed, an extension of bidding period should be granted, and the reasons for delay be stated. The revised dates for receipt or opening of bids shall be communicated to all bidders, and addenda should be issued as mentioned above.

8.4.4 Bid Security or Bid Bond

As a standard practice, a bid bond is usually required by the Owner from the bidders at the time of submitting bid. Bid bonds are also known as bid securities and call deposit. Like other bonds, the form of bid bond is also included in the bid documents. The amount of bid bond required is mentioned in the bid documents either in a specific amount or as a percentage of contract bid price.

A bid bond is helpful in protecting the Owner against risks associated with the bidding process. Bid bonds can demonstrate that a bidder is reliable and genuine and has the ability to complete the project under bid. A bid bond does not, however, guarantee construction of the project but insures against the bidder's failure to adhere to the bid instructions and conditions. These bonds serve as a form of liquidated damages and can be used to provide incentive against bid withdrawal [3].

A bid bond may be in cash, letter of credit, government bonds, or letter of guarantee; however, the sureties must be satisfactory to the Owner. After the bids are open, the bond is released either after a specified period or earlier if one of the bidders has been accepted, and he submits the performance security. In the event the accepted bidder fails to provide a performance security within the specified period or refuses to enter the contract, his bid bond will be forfeited, and the second lowest bidder will be contacted.

8.4.5 Bid Forms

The bid form is a project-specific form and is typically required by any bidder responding to a bidding process for construction or rehabilitation projects. The layout of the form and the information required is provided by the bid-calling authority. The purpose of this form is to provide a one or two page summary of the essential information related to the bid.

The bid form is usually divided into four sections: contact information of the bidder, project brief, bid details, and essential notes. The contact information of bidder includes the bidding firm's legal and operating name, head office address, primary and secondary contacts, as well as telephone, email, and fax numbers. The first section of a bid form also includes the bid-calling authority's name and address

The project details usually include the bid number, brief about the proposed project including addendum number (s), total completion, and substantial completion dates of the project. Details about the bid are typically the total dollar value and any specific exclusions. For example, design and construction of a wastewater pumping station may include the total dollar value of the project, plus a statement that the costs for environmental assessment and approval are not included. Essential notes include information that forms part of the bid, e.g., a note saying that "all supplements to the bid form required under the bid documents form an integral part of this bid."

CCDC 23 provides sample model bid forms for stipulated and unit price contracts with a recommendation that the bid-calling authorities use these forms to create their own, project-specific bid forms [1].

8.4.6 Supplements to Bid Forms

Supplements to the bid form are additional documents that are required as part of the bid submission. This may include:

(i) Bid security form
(ii) List of Subcontractors
(iii) Bidders qualification forms
(iv) Alternative prices
(v) Itemized prices
(vi) Schedule of unit prices

8.5 Bidding Procedures

The type of acquisition depends on the type of product or service requested and the anticipated cost. Most commonly there are two tendering procedures in practice: competitive and negotiated; however, electronic bidding is also being used by many organizations now-a-days. Under the first approach, the Owner invites bids from the contractors to construct the works. The contractors compete with one another by submitting their proposals to the Owner. The contract is concluded on the basis of the bid selected by the Owner through formal bidding procedures. Under the second approach, the Owner negotiates the contract with the contractors selected by him without formal bidding procedures.

Competitive bidding can be achieved by three different modes: open, single-stage or two-stage bidding.

8.5.1 Competitive Bidding

8.5.1.1 Open Bidding

Under this procedure, the details of the proposed project are advertised in the press. The advertisement should carry the details as already mentioned in the "invitation to bidders." Contractors who are interested in bidding for the work can apply and collect the documentation by paying a small fee.

Open bidding can bring many unknown Contractors, who are generally not vetted, before the bids are submitted. Factors other than price must therefore be taken into account, while evaluating bids are submitted. It is generally accepted that a lowest price bid will be obtained by the use of this method. There is, of course, no obligation on the part of the Owner to accept the lowest or any of the bids.

Because of the indiscriminate nature of open bidding, contractors can be awarded work for which they are not properly equipped, in terms of either resources or experience. Although the Owner is not bound to accept the lowest bid, acceptance of a higher bid is rare due to general practice and departmental requirements. Because of the problems associated with open bidding, its use has been declining in recent years. For major contracts, it is not advisable.

8.5.1.2 Single-Stage Selective Bidding

This is the most popular method of awarding construction contracts. The purpose of this procedure is to restrict the number of bidders being invited to submit bids. Basically, it involves a pre-selecting or pre-qualifying of a limited number of satisfactory contractors to bid for a project. The firm to submit the lowest bid is then awarded the contract.

This bid procedure is also called one-envelope bidding, because, sometimes, the bid calling authority requires both pricing and technical proposal together (at the same time). In that case, the bidders submit bids in one envelope, containing both the price and the technical proposal, which are opened in public at the date and time advised in the bidding documents. The bids are evaluated, and, following review and approval, the contract is awarded to the lowest evaluated bidder.

This method of bidding is appropriate for almost any type of construction project. In order to short list the contractors, the suitability can be judged on the following grounds:

(a) The firm's structure and organization setup
(b) The firm's previous experience for a similar nature of projects completed satisfactorily
(c) The firm's reputation indicating responsibility and reliability
(d) The available resources and capability to handle such projects
(e) The firm's financial position
(f) Adequacy of personnel and equipment
(g) Frequency of subcontracting the main works

8.5.1.3 Two-Stage Selective Bidding

This process of bidding is being used on large or complex projects. During the first stage of the system, bidders submit their Technical Proposals except price, in accordance with the Contract Documents. The Technical Proposals are opened at the date and time advised in the bidding documents. The Technical Proposals are evaluated and discussed with the bidders. The bidders are informed about any deficiencies and unsatisfactory technical features recorded in the submitted bids. The bidders are allowed to revise or adjust their Technical Proposals to meet the requirements of the bid-calling authority.

The objective of this exercise is to make sure that all Technical Proposals conform to the same acceptable technical standard and meet the technical solution required by the bid-calling authority. Bids of those bidders who are unable or unwilling to bring their bids to conform to the acceptable technical standard may be rejected as deficient bids.

After the evaluation of technical proposals has been approved by the bid-calling authority, the second stage is to invite bidders to submit price proposals and revised technical proposals in compliance with the acceptable technical standard. The revised technical proposals and price proposals are opened in public at a date and time advised by the bid-calling authority.

The price proposals and revised technical proposals are evaluated, and the contract is awarded to the bidder whose bid has been determined to be the lowest evaluated substantially responsive bid.

8.5.2 Negotiated Bidding

Instead of inviting competitive bidding, the Owner contacts one or more contractors, which he judges to be capable of constructing the works. The Owner informs them about the requirements and requests offers. Normally, this type of noncompetitive, negotiated contract is limited to privately financed work because competitive bidding is required for government/public projects except under extraordinary or unusual circumstances.

Documents describing the scope of work including the contractual terms required by the Owner are submitted to the contractors. No formalities are usually required for making or evaluating the offers, or for negotiating the contract. In certain circumstances the Owner may combine the bidding and negotiating approaches.

A major reason for using negotiated bidding is the flexibility of this type of pricing arrangement, particularly for projects of large size and great complexity or for projects which substantially duplicate previous facilities sponsored by the Owner. The Owner may value the expertise and integrity of a particular Contractor who has a good reputation or has worked successfully for the Owner in the past. If it becomes necessary to meet a dead line for completion of the project, the construction of a project may proceed without waiting for the completion of the detailed plans and specifications with a Contractor that the Owner can trust. However, the Owner's staff must be highly knowledgeable and competent in evaluating Contractor proposals and monitoring subsequent performance.

8.5.3 *Electronic Bidding*

Electronic bidding systems are being used now-a-days by many organizations throughout the world. It is basically a fully automated web-based system using an e-commerce concept. The system is created in such a way that the data is stored in the bid-calling authority's server, which has reliable data security. In order to provide additional security, the sensitive fields such as price and password are kept encrypted.

Electronic systems promise greater accuracy and efficiency, potentially saving time and money. Costs are reduced for paper, filling and handling, postage, and data storage. The system enables the bidders to view and respond to all their bids before bid-closing date. Prospective bidders are first registered by the service provider or bid-calling authority, who are then given a secure access key with which to submit their bids. The system allows the bidders to download bid information from the required web site and fill in prices interactively. Item extensions and section and proposal totals are calculated and displayed immediately. The software alerts the bidders for any omissions and incorrect alternates and prevents them from making calculation mistakes.

With an electronic system, the bid-calling authority, however, has to ensure that the bids were properly received and that the bidders have access to the bid results promptly after bid closing. Additionally, they have to ensure the security and confidentiality of bid submissions and bid modifications. CCDC-23 [1] suggests that the bid-calling authority should provide for an alternative backup process in case of unforeseen system malfunctions, disruptions, or other problems.

Most widely online services for electronic bidding used in Canada are MERX and Biddingo. As per MERX, legislation in Canada and the United States ensures that electronic submissions have the same binding effect as traditional paper bids.

8.6 Receiving and Opening of Bids

Bids received should be marked with time and date of arrival on the outer envelope. Bids should be kept secure and unopened, better to be deposited in a bid box until the date and time of the opening. Bids should only be removed from the box in the presence of the opening committee and at the time stated in the instructions to bidders. Bidders should obtain a receipt upon delivery of the bids to the Owner or Consultant which records the date and time of delivery.

Once the bids are received within the specified period, they are opened at the date and time already fixed and as announced in the invitation to bid. Prior to the opening of bids, the seals of each envelops are checked. The bids should be opened either in public or in the presence of bidders involved. For each opened bid the following information should be read out at the bid opening:

1. Bidder's name
2. Date and time of receipt
3. Bid price
4. Modification and bid withdrawal, if any
5. Required bid bond

The names of disqualified bidders due to late or non-receipt are also announced.

CCA 29 [4] suggest that if the tenders are not opened in the presence of bidders, a list of bidders including bid price should be sent to all bidders immediately following the bid opening.

8.7 Bid Evaluation

The main purpose of bid evaluation is to determine the best responsive and compliant bid, in accordance with the evaluation and selection methodology specified in the bidding documents. Fair, accurate, and transparent evaluation of bids is an important aspect of procurement process. Evaluation of bids should be completed as soon as practicable after the opening of bids to allow for the award of the contract prior to the expiry of the bid validity period. Any irregularity that may cause the bid to be considered non-compliant must be rejected as per law of competitive bidding. The following section provides a summary of the bid evaluation process and the associated components:

8.7.1 Administrative Evaluation

8.7.1.1 Incomplete Bids

Bids should first be evaluated to check for omissions of any documents or failure. A bid is incomplete and non-compliant, if the bidder has neglected to include significant information and would be subject to disqualification. For example, in a unit price bid, if the bidder has omitted a unit price, the bid should be disqualified. Permitting the bidder to provide the missing unit price at this stage would give the bidder an unfair advantage over other bidders.

8.7.1.2 Non-responsive Bid

If a bid fails to satisfy the mandatory terms, conditions, and instructions contained in bid invitation, the bid would be considered non-responsive and will be subject to disqualification being non-compliant. Examples of non-responsive bids are missing critical pricing information or missing bid security. Also if the bid is qualified in some way, the bid is to be considered non-responsive. For example, if the bidder changes the requirement "Supply and install 16 *metal* windows" to "Supply and install 16 *wood* windows," the bid is non-responsive [5].

8.7.1.3 Qualified Bids

Qualified bids are those that contain qualifications. For example, the lowest bidder is not prepared to provide a heating system to meet the specified requirements but proposes something of substantially lower quality and performance. Qualified bid will also be treated as non-compliant and cannot be accepted.

Examples of qualified bids are [5]:

1. Bidder does not accept the terms of payment.
2. Bidders propose different insurance coverage.
3. Bidders propose different guarantees than those specified.
4. Bidder is not prepared to provide a certain component as specified in the bid invitation but proposes something of different quality and performance.

8.7.1.4 Identical Low Bids or Tied Bids

Traditionally and most commonly construction contracts are awarded to the company with the lowest-priced confirming bid. However, there are occasions when the traditional bidding method may not result in the selection of the most suitable Contractor for the project, for example, when identical low bids are received or when a contract for long-term maintenance of a structure is required. In such situations, an alternative bid evaluation process using weighted criteria or best value criteria is used to determine the bid that offers the best value. The Treasury Board of Canada suggests under subsection 10.8.17 of Contracting Policy that if identical low valid bids are received, the contract should be awarded on the basis of best value [6].

Best value criteria is a process based on competitive bidding that awards the contract to the bidder offering the best value, taking into consideration the best performance record in terms of time, cost, and value for money. Weighted criteria method of bid evaluation must be decided before bidding and should be included in the bid documents forming part of bid assessment process.

The best value criteria for performance can be based on the following factors:

1.	Overall past performance on previous projects	15%
2.	Quality management system on previous projects	10%
3.	Claims reputation on previous projects	10%
4.	Safety record on previous projects	10%
5.	Technical capabilities	10%
6.	Financial capabilities	10%

7.	Own manpower and construction plant capabilities	10%
8.	Pricing strategy	10%
9.	Abilities to assess and minimize risk	5%
10.	Methodology	10%
	Total	100%

Based on the above criteria, the sample bid evaluation sheet could be as shown in Table 8.1 below assuming the three lowest close bids: Bid A $10,200,000.00, Bid B $10,190,000.00, and Bid C $10,205, 000.00.

If price is considered as the decision criteria, then Contractor B needs to be selected being the lowest quote. However, if best value criteria is the decision factor, then based on score, Contractor C would be awarded.

8.7.1.5 One Responsive Bid

When only one responsive bid is received in response to a competitive bid solicitation, the bid amount is reasonable and within budget, and delivery of the requirements can be made in time, the sole bid can be accepted.

If the bid-calling authority is not satisfied that the bid represents fair value, price justification must be requested from the bidder. If this does not show that the price is fair and reasonable, the bid-calling authority may consider negotiating or canceling and inviting new bids.

8.7.2 Technical Evaluation

All bids must be checked for substantial responsiveness to the technical requirements of the bidding documents. Technical evaluation is usually based on mandatory criteria and rated/points criteria.

8.7.2.1 Technical Mandatory Criteria

Mandatory criteria are judged on a simple pass/fail basis. Bids are considered non-responsive if they fail to meet any of the mandatory criteria. The reasons for declaring a bid non-responsive are documented. For example, if a bidder fails to quote for a major item in the bid package, it will be rejected and will not be assessed on the basis of a point rating.

Table 8.1 Best value criteria for tied bids

Criteria	Weight (%)	Bid A		Bid B		Bid C	
		Rating	Score (weighted points)	Rating	Score (weighted points)	Rating	Score (weighted points)
1	15	1.5	0.22	1.5	0.22	2.5	0.37
2	10	2	0.2	2.5	0.25	2.5	0.25
3	10	2	0.20	1.5	0.15	2.5	0.25
4	10	2.5	0.25	2	0.20	2	0.20
5	10	3	0.30	2.5	0.25	3	0.30
6	10	3	0.30	2.5	0.25	3	0.30
7	10	2	0.20	2	0.20	3	0.30
8	10	1.5	0.15	2	0.20	2.5	0.25
9	5	1	0.05	1	0.05	2.5	0.12
10	10	1.5	0.15	1	0.05	2.5	0.25
Total	100		2.02		1.82		2.59

Rating criteria: 1 = poor, 1.5 = average, 2 = satisfactory (meets requirements), 2.5 = very good, 3 = outstanding

8.7.2.2 Technical Rated Criteria

Only bids that pass the mandatory criteria will be subject to point rating, as applicable. Rated criteria are used to assess various elements of the technical bid so that the relative merits of each bid can be determined. The point-rated method is described above, and the same parameters will be applicable; however, evaluating factors can be changed. For technical evaluation, the following factors can be assessed:

1. Plans and schedules
2. Conformity with specifications and drawings
3. Construction methods and sequence
4. Construction plant and equipment
5. Proposed subcontractors
6. Technical and administrative personnel
7. Skilled and unskilled manpower
8. Bill of quantities and unit rates

8.7.2.3 Unbalanced Bids

A bid is considered unbalanced if the unit rates quoted are either excessive or below the reasonable cost analysis value. There are two types of unbalance bids: mathematically and materially.

1. A mathematically unbalanced bid is that which contains lump sum or unit bid items which do not reflect reasonable actual costs including a reasonable share of bidder's anticipated project overhead costs, profit, and other indirect costs.
2. A materially unbalanced bid is that which creates a reasonable doubt that award to the bidder submitting a mathematically unbalanced bid will result in the lowest overall cost to the bid-calling authority.

There are many reasons for unbalancing a bid. One reason is to get more money at the beginning of the project by raising the price on certain bid items. This is called "front loading" the contract. Another reason is to maximize the profit by overpricing those bid items which bidder assumes will be used in greater quantities than estimated and underpricing those items which will be used in lesser quantities.

Sometimes, an unreasonably low bid price may be due to a bidder's misunderstanding of the specification, and accordingly, this bidder may incur substantial financial losses and fail to complete the work satisfactory if awarded the contract. To assess the reasonableness of a bid price, it is necessary to consider all circumstances affecting the bid.

Many organizations accept a bid that is found to be mathematically unbalanced and reject a low bid that is found to be both mathematically and materially unbalanced. A bid-calling authority must clearly specify in the bid documents the policy for evaluation of unbalanced bids.

8.7.3 *Financial Evaluation*

Bidders must demonstrate the capacity to meet the financial requirements of the contract. To determine the bidder's financial capability, the financial information detailed below can be considered:

1. Audited financial statements for last 3 years, including the balance sheet, the statement of retained earnings, the income statement, and any notes to the statements.
2. Value of projects completed during last 3 years and projects in hand.
3. Reference letters from financial institutes.
4. Reference letter from insurance company and bonding company.
5. WSIB (Work Place Safety and Insurance Board) clearance certificate.
6. Details of any outstanding claims or litigation against the company.
7. If the bidder is a joint venture, then each member of the joint venture must provide the required financial information.

If errors, omissions, or inconsistencies are apparent, within the bids, a meeting should be arranged with the lowest two or three bidders, to clarify the position and to agree how to deal with the points in the event of an award. At such meetings, bidders should not be permitted to change the substance of their bids. If differences are not resolved, the particular bid should be treated as unresponsive, and no further consideration should be given to that bid.

Finally, a summary comparison of all the bids with the Consultant's estimate should be made. A recommendation should be made stating which bid is recommended for acceptance or stating that none should be accepted. Only one bid should be recommended for acceptance.

8.8 Award of Contract

After completion of the process of tender evaluation, the Consultant makes a recommendation as mentioned above, to the Owner on the award of the contract to the successful bidder. In most cases after agreeing with the recommendations of the Consultant, the Owner issues a letter of intent on the basis of "subject to contract." Such a letter should state the conditions, usually related to the submission of a performance bond and insurance certificate that must be met before the award can be formalized. In practice, the letter of intent has no legal effect. No binding contract will come to in effect until the formal "letter of acceptance" is issued.

Sometimes a letter of intent gives instructions to the potential Contractor to take some action, such as to order materials and plants or to carry out limited work. In this case, the letter of intent should clearly mention how and to what extent, the Contractor will be paid for his work done, because sometimes for some reason, the parties may not enter into a contract.

Once the successful contractor provides the performance security, required sureties, insurance, and any other requirements on letter of intent (if any), signing of the contract should be arranged. After the signing, the Owner issues a "letter of acceptance" which, together with the bidder's bid form, comprises a binding contract between the two parties, valid from the date of issue of letter.

References and Further Reading

1. CCDC-23. (2005). Canadian Construction Document Committee. A guide to calling bids and awarding construction contracts.
2. The FIDIC. (1999). Conditions of contract for construction – Red book. Author FIDIC – International Federation of Consulting Engineers.
3. Course Notes (2005). From Engineering Ethics and Law attended by Author. Conducted by EPIC Canada (Educational Program Innovation Center).
4. CCA 29 (Canadian Construction Association-1995). A guide on standard contracting and bidding procedures.
5. Northwest Territories Public Works & Services Canada- Procurement Guidelines (2013, September). Chapter 3.8; Tender evaluation.
6. Treasury Board of Canada- Contracting Policy. Date modified 9 Oct 2013.

Chapter 9
Construction Risk Analysis and Management

9.1 Introduction

Risks associated with construction are numerous, and each of the four main project objectives of scope, time, cost, and quality remains subject to risk, and their effects are considered throughout the project. Construction projects can be unpredictable. Risks start appearing as soon as it is decided to initiate a project and continue to increase as the project moves further.

Construction projects can be extremely complex with many uncertainties. In any complex project, there are many things that can go wrong. Risks are therefore inevitable and cannot be eliminated; however, they can be mitigated or distributed with better preplanning. Construction contracts basically carry risks for both the Contractor and the Owner. Unfortunately, some project owners try to allocate the majority of risks to the Contractor satisfying their needs. Whereas contractors consider that they can minimize risk through early completion of the project for less than the contract price, to maximize profits.

In order to achieve project objectives, managing risks in construction projects has been recognized as a very important process. Good risk management practices direct that the distribution of project risks between the various parties shall be made clearly, sensibly, and equitably. Unfortunately, there is a general failure between the construction industry to handle risk issues properly, which mostly results in excessive claims and disputes within the parties. It is therefore important to be prepared for risk occurrences and to handle them timely, through good management and foresight.

9.2 Definition of the Risk

Various definitions have been provided by various professionals. Darnell and Preston (2010) defines risk as "possibility of loss or injury." Barber (2005) defines risk as "a threat to project success, where the final impact upon project success is

A. Surahyo, *Understanding Construction Contracts*,
DOI 10.1007/978-3-319-66685-3_9

not certain". The National Association of Surety Bond Producers [1] defines construction risk as "any exposure to possible loss".

Risk is basically the product of severity of hazard and the probability of occurrence. Because every construction project is unique, each offers a multitude of different risks. To ensure the success of an undertaking, the Owner initiating a construction project must be able to recognize, assess, and handle these risks.

9.3 Risk Management

Risk management is concerned with planning, identifying the risks, assessing their likelihood, and deciding how best to manage the project efficiently in the light of this information. An effective risk management method can help properly in identifying risks and how to manage these risks in different stages of a project. Effective risk management increases the likelihood of project success increasing the likelihood of finishing the project on time, within budget and meeting stakeholders' performance expectations.

For effective management of risks, efforts are required at the early stages of the project to identify all relevant risks by all team members. A key to the success of any construction project is having the right Project Manager, with the right project management model, team, and supporting resources. A Project Manager must be able to recognize and identify the root cause of the risks and mitigate them well. The process of managing contractual risks falls into five main stages.

9.3.1 Risk Management Planning

Risk management planning is the process in which it is decided and worked out how to approach and conduct the risk management activities for a project. Risk management planning should be carried out during the early stages of the project, i.e., at the project planning stage. Risk planning will help earlier identification of high risks, limiting inappropriate changes, rework, cost overruns, time overruns, etc.

To develop a risk management plan, the project team should conduct internal meetings with all related staff. Mainly the risk management plan should include:

1. Procedures, means, and methods to be used to perform risk management on the project
2. Roles and responsibilities (complete with details of who will do what)
3. Budget and time to deal with potential risks
4. A process for identifying risks including risk monitoring procedures

9.3.2 Identifying the Risk

The second step in dealing with construction risks is to develop a method for the identification and classification of individual risks related to the particular project. Every construction project carries various different types of risks; however, some risks are common to most of the projects. Examples include differing site conditions, inclement weather, contractor reliability, and the risk of maintaining adequate funds.

A common procedure for risk identification and classification is the development of a risk checklist. For preparing a risk checklist, the past record of risk management for the completed projects is reviewed. This technique allows the user to list common project risks and then to add those risks which are specific to the project in hand.

9.3.2.1 Organizing Checklist

There are several approaches for organizing a risk checklist into a suitable format. One approach proposes that risks should be organized in terms of the nature of the risk itself, like known risks and unknown risks. A second approach proposes to classify the risks upon their effect on the project, like cost risks, schedule risks, or quality risks.

As such, risk is defined in terms of an event, its probability, and the cost and time involved; a suitable way of organizing a construction risk checklist is to divide risks into two broad categories: design and construction risks and financial risks [2]. These risks are most related to the construction projects. Financial risks start affecting the project from the earlier stages of planning and feasibility phases, whereas, design and construction risks appear throughout the life cycle of a project and especially during the construction period.

Most types of risk that occur on a construction project can be summarized using the following checklist, Table 9.1, and can be tailored to a specific project in hand.

9.3.3 Analyzing the Risk

The third step in managing risks is analyzing the risks, i.e., assessing the probability of each risk occurring and severity or impact on main project objectives: cost, time, and quality. Risk assessment methods are formal processes for employing common sense and creative lateral thinking to visualize problems and workout ways to reduce or avoid the likelihood of their occurrence. They depend on good communication, good documentation of ideas and discussions, and agreement about how the whole process will be handled. Risk assessment needs to begin at an early stage and be continuously updated as the project progresses and details are clarified. Risk

Table 9.1 Risk category checklist

Risk category	Risk factor
1. Funding and payment	Lack of financial resources/funding
	Delay in payments
	Delay in settling changes & claims
	Inflation
	Owner or contractor's insolvency
2. Planning and designing	Technical feasibility
	Economic viability
	Inadequate statement of work (scope)
	Project complexity
	Sole source of material or services providers
	Constructability
	Program of works
	Design completeness and standards
	Inadequate selection of contract types (e.g., lump sum, unit price, cost plus, etc.)
	Inadequate selection of contract delivery methods (e.g., traditional, design and build, management, etc.)
3. Construction	Shortage of staff, labor, plant, or materials
	Labor productivity
	Strikes and unions
	Work ethics
	Wage scales
	Labor disputes
	Delay in possession of site
	Underground conditions or differing site conditions (soil conditions, water, utilities, archeological findings)
	Inclement weather
	Hazardous wastes
	Noise, fume, and dust
	Defective materials and workmanship
	Contractor reliability (e.g., capacity, capability, etc.)
	Subcontractors inefficiency
	Delayed drawings or instructions
	Errors in design and drawings
	Incomplete and inefficient supervisory staff
	Poor planning and management
	Poor communication and coordination
	Scope changes and claims
	Too much Owner involvement
	Acts of God (e.g. storms, earthquake, floods, etc.)
	Vandalism
	Accidents
	Third-party claims

(continued)

Table 9.1 (continued)

Risk category	Risk factor
4. Regulatory conditions	Licenses, permits, and approvals
	Environmental regulations and requirements
	Taxes and duties
	Health and safety regulations

assessment is an important task, which must be carried out successfully by using reasonable methods of measurement.

Risk assessment should commence in conjunction with the identified risk items in the checklist described earlier. Using the checklist allows the division of risk items according to categories which contributes to a better understanding of how these uncertainties function and affect the project. The risk assessment process involves itemizing the risks into a ranking that will place them in order of importance. To do this, every item in the checklist should be earmarked as high, moderate, or low risks (Table 9.3). For example, if an individual project involves major underground construction, then the risks associated with this activity will become very important and will require extra attention. The checklist can be examined for every project and filled in so as to reflect specific project characteristics. This method is called Qualitative risk analysis.

The second risk analysis method is Quantitative analysis which means numerical analysis of the probability and impact of the project risks. This method is initiated for the highest risks from qualitative risk analysis and is not recommended for a low priority risks. Under Quantitative analysis the most common method of assessing risk is to prepare a fault tree of possible hazards. If the hazards are acceptable, the risk is calculated by using the concept: risk = impact of hazard x probability of occurrence [3]. If this calculated risk is accepted, it is then distributed among the required parties. However, if the hazards are unacceptable, the activity is either rejected or a method is found to mitigate the hazards and carry on further process.

This method can best be understood by an example. Suppose a sewer treatment plant plan to install fire fighting system in the plant. They have come up with the following two options to decide whether to proceed with this item of work or not.

Example 9.1

Scenario 1: Fire fighting system required	Scenario 2: Fire fighting system not required
Set-up cost = $100,000.00	Set-up cost = $0.00
Issues, problems, maintenance = 20% probability and $50,000.00 impact	If not installed then risk chances = 60% probability and $300,000.00 impact
Analysis for this case = 20% × $50,000 + $100,000.00 = $110,000.00	Analysis for this case = 60% × $300,000.00 = $180,000.00

At initial stage looking at the set-up cost against scenario 1, plant authorities might think that there is no need to invest $100, 000.00 for this item in comparison to scenario 2 where the set-up cost is zero, but the analysis proves differently.

The effectiveness of this technique depends on the ability of those engaged to identify all possible events. It is worthwhile to mention that risks having very high probabilities and very low impacts or vice versa are usually not considered important in the construction projects. In fact, it is the combination of probability and impact that matters. Other Quantitative risk analysis methods include Monte Carlo simulation, cost-risk analysis, expected monetary value analysis, etc.

9.3.4 Responding and Allocating Risk

The fourth step is to decide which party is best equipped to manage, control, and assume the risk. Risk response involves procedures to either eliminate a risk before it occurs by proactive approach or mitigate its probability and impact as much as possible. Some risks may be insured, as within most of the standard forms of the contracts, risk is primarily allocated between the parties through indemnity and insurance requirement provisions.

Since funding is provided by the Owner, it is his privilege to assign responsibilities. He has the opportunity to reduce the total project cost through effective allocation of related construction risks. However, as mentioned earlier, the distribution of risks between the parties should be carried out sensibly and equitably because misallocation of risks is not cost-effective. Placing an inequitable risk share on the Contractor should be avoided as it promotes negative working relationships and increases disputes.

Proper and reasonable risk allocation improves efficiency, reduces costs and disputes, and promotes project goals. In the absence of uncertainties imposed by unfairly allocated risks, contractors can avoid the addition of cost contingencies in the pricing of project bids and estimates. By adopting fair risk allocation, the Owner can expect that their projects will have fewer claims, reduced costs, and timely completion. The most common options of responding and allocating risks are risk acceptance, risk sharing, risk avoidance, risk reduction or mitigation, and risk transfer.

Riskacceptance The most appropriate method to allocate risk is to have the party who controls the risk should accept/retain the risk. For example, a Contractor who is in charge of the construction works, risks arising out of his operations should be allocated to him. Similarly, as the Designer is in control of the design component, he should be allocated the risks pertaining to design, e.g., design errors. Similarly, site selection is the Owner's responsibility; hence, risks pertaining to site issues such as access, permits, and acquisition should be allocated to the Owner. In the same way, if the risk of fire is to be dealt with by an indemnity from an insurer, the Contractor who is responsible will be tasked with obtaining the insurance coverage and will be considered as having control over the policy; thus, the risk is allocated to him.

Table 9.2 Risk matrix example showing risk and responsibilities of the parties

Activity	Owner	Consultant	Contractor
Possession of site	X		
Financial resources	X		
Errors in design and drawings		X	
Weather (normal)			X
Inclement weather	X		
Acts of God	X		
Schedule			X
Productivity			X
Adequate manpower and so on			X

Under standard forms of contracts, the Owner will also retain risks that are out of the control of the Contractor, such as acts of God, war, riots, nuclear explosives, strikes, etc. Risk acceptance may be made with contingency. A typical contingency sum will be allowed to the project to address these risks. Contingency is a sum of money or period of time set aside from the general construction funds to pay for losses that actually occur.

The best way to identify the risk assumed by the various parties is to develop a risk matrix. Each party should refer to the matrix and evaluate the extent of risk they are willing to undertake. Parties also should ascertain who is to bear the responsibility for other risks not listed in the matrix. Table 9.2, indicates an example of a risk matrix showing some of the risks and the responsibilities of the parties involved.

Risk sharing Sometimes it is not possible for one party to control a specific risk. The suitable procedure is that it may be then shared and managed by dividing it among two or more parties so as to manage the portion that they are best able to control individually. An excellent example of risk sharing is the development of a joint venture by contractors [2]. A joint venture is the result of the unification of two or more contracting firms to build a single project. Hence all partners share in construction risk in the same way they share in any profit and loss.

Another risk sharing system can be found in the guaranteed maximum price contract. In this contract type, the Owner and the Contractor agree to a project cost guaranteed by the Contractor as maximum. With a guaranteed maximum price contract, amounts below the maximum are typically shared between the Owner and the Contractor, while the Contractor is responsible for costs above the maximum.

Some individual risks are also shared between the Owner and Contractor under standard forms of contract. For example, the risk resulting from the provision regarding "adjustment for changes in legislation" is shared between the Owner and Contractor both under FIDIC Conditions of Contract (Red Book) [4]. FIDIC Sub-clause 13.7 refers to increase or decrease in cost due to changes in Legislation. Accordingly, if a Contractor suffers delay or additional costs, the Owner has to share the burden. However, if there is decrease in cost, then the Owner can submit a claim for a credit.

Table 9.3 Risk response matrix

Risk event	Probability of occurrence			Magnitude of impact			Risk response
Name	High	Medium	Low	High	Medium	No action	Type of action
Dewatering	X			X			Risk allocated to the contractor
Addition of internal walls (scope changes)		X			X		Risk allocated to the contractor
Indemnities	X			X			Shared

Risk avoidance Avoiding a potential risk will reduce a risk's probability to zero. After the risks are identified and analyzed, some risks may be found unacceptable, e.g., deficiency of funds. Withdrawing a bid by the Contractor or the Owner and not proceeding with the project are two examples of total avoidance of risks. In such case the Owner may redefine (i.e., reduce) the scope of work or even abandon the project for a period of time until the funds are made available to avoid the risk. Some risks that arise early in the project can be avoided by clarifying requirements, obtaining proper information, improving communication, or acquiring expertise.

In addition to risks between the Contractor and Owner, consultants are also subject to risks. Consultants can avoid risks by ensuring that the professional responsibilities are clear and unambiguous.

Risk reduction / mitigation Risk mitigation involves reducing the probability or impact of the risk to an acceptable level. Actions are taken before the risk is triggered to minimize its probability of occurrence and impact. Taking early action to reduce the probability or impact of a project risk is usually more effective than trying to rectify the damage after the risk has occurred.

The most common risk mitigation efforts include having strong project control, conducting constructability reviews, expediting variations and claims, avoiding late issue of drawings, use of standard items or software, using alternative design and specifications, etc. In short, a well-managed project will have fewer risks.

Risk transfer Risk transference involves shifting the negative impact of a risk threat, along with ownership of response to a third party. Risk transferring also eliminates the risk from impacting the project. However, when risks are transferred to another party, a premium is paid to compensate the third party to take on the risk.

All parties involved in a construction project can protect their interests via insurance but must accept that not all risks are insurable as described earlier. Many of the standard form contracts insist on certain types of insurance. Insurable risks are generally allocated by contract to the party best able to control, manage, and insure the risk. Standard insurable risks are for works such as, materials, plants and Contractor's equipment, damage to person and property (third-party insurance), etc.

The consultants usually obtain "Professional Indemnity Insurance" so as to cover themselves and their clients against the risk of failure to perform their duties with the required level and skills.

The insurance and surety industries provide financial security for parties that incurred legal obligations. Accordingly, forms of financial security and guarantee are available to transfer certain types of risk. For example, the risk of nonperformance on a contract can be addressed by a financial guarantee.

Once the risk assessment and risk response actions are finalized, the Risk Response Matrix can be prepared for proper record and further review during the risk monitoring and control process. Table 9.3 provides an example of a Risk Response Matrix with some of the risk examples.

9.3.5 Risk Monitoring and Control

Risk monitoring and control is the process of tracking all the identified risks, analyzing and reviewing these risks, and identifying and planning for newly arising risks during the life of the project. The progress and performance of risk responses are evaluated, and risk status is updated periodically.

The main objective of the risk monitoring process is to establish a cost performance and schedule management system. The system should be designed to provide early warning of potential problems to allow management action. As the risk approaches, the risk strategies are reviewed for appropriate actions, like choosing alternative strategies, executing a contingency plan, taking corrective action, etc.

As each risk occurs and is dealt with or is avoided, these changes must be documented. Good documentation insures that risks of this type will be handled in a more effective way than before and that the next project manager will benefit from "lessons learned."

9.4 Standard Forms of Contract

In order to achieve fair distribution of risks, standard forms of contract should be used, which carries appropriate clauses for allocation of risks. For example, clause 17 of FIDIC [4] allocates responsibility between the Contractor and the Employer for damage, to the works. According to this clause, the Contractor is to bear the cost of rectification of loss or damage, which arises from any cause, other than those, which are described as "Employer's Risk". Where damage is caused by an Employer's Risk, the cost is born by the Employer.

Generally, risks that are severe and outside the control of the Contractor, like war, earthquake, riots, nuclear explosives, etc., should be retained by the Employer as is specified by the FIDIC [4] in abovementioned clause.

Construction industry standard forms are produced by construction industry experts and construction law practitioners over a period of many years after receiving suggestions and feedback from owners, contractors, consultants, sureties, and insurers and are also being updated from time to time. Despite all these efforts, standard forms should not be used without modifications as they are drafted for broad applicability. Standard form contracts cannot account for all specific legal and financial terms that the parties need to insert in their agreements. The activity of allocating risks should be carefully evaluated by an insurance professional, contracts specialist, legal counsel, or knowledgeable design professional to assess specific or unique risks and exposures.

References and Further Reading

1. National Association of Surety Bond Producers – SuretyLearn.Org. Washington, DC, USA.
2. Touran A., Bolster P. J., & Thayer S. W. (1994). Northern University, Boston. Risk assessment in fixed guide way transit system construction. A report prepared for U.S. Department of Transportation, Washington, DC.
3. Bunni, N. G. (1986). FIDIC's view of design liability- Construction, Insurance & law. A discussion paper. Published by FIDIC.
4. The FIDIC. (1999). *Conditions of contract for construction – (Red book).* Author FIDIC.

Chapter 10
Construction Insurance and Construction Bonds

10.1 Construction Insurance

10.1.1 Introduction

Construction projects are a complex and risky endeavor from start to finish. Construction work, by nature, is hazardous, and accidents are frequent and sometimes severe. The annual number of fatalities, personal injuries, and property damage in the construction industry is higher relative to most other industries. Parties therefore mostly turn to construction insurance policies to protect themselves against such risks and to help ensure that the benefits of undertaking a project are not outweighed by its potential costs.

There are provisions under standard forms of contracts that the Contractor shall be liable to the Owner for certain kinds of damages, however, the Owner's right to sue the Contractor will be fruitful only when the Contractor would be able to pay any damages awarded. Thus the Owner can be protected if the Contractor carries either liability and loss insurances or a performance guarantee.

Construction contracts typically require the Contractor to assume the Owner's and Consultant's legal liability for construction accidents or to provide insurance for the Owner's direct protection. The Contractor's insurance program normally includes liability insurance coverage and loss insurance coverage (usually for works). Unfortunately, insurance requirements often are viewed as routine paper work and do not receive proper consideration until an accident or other loss occurs. The parties then look to their insurance companies for coverage, sometimes only to find that protection is not there. A basic understanding, therefore, of insurance procedures is vital for the parties involved in the construction business.

A. Surahyo, *Understanding Construction Contracts*,
DOI 10.1007/978-3-319-66685-3_10

10.1.2 Insurance Policy

An insurance policy is a conditional contract (legal document) between the insurer and the insured (known as the policy holder), under which the insurer promises, for a consideration, to assume financial responsibility for a specified loss or liability.

Insurance can be defined as "the equitable financial contribution of many for the benefit of an individual who has suffered loss" [1]. With these policies, one party (the insurer) agrees to pay those sums that the other party (the insured) becomes legally obligated to pay as damages on occurrences of any specified loss to which the insurance applies. The insurer also agrees to defend any suit seeking damages.

As the terms and conditions may differ, it is dangerous to generalize or make assumptions about insurance policies. For example, one cannot assume that all Contractor's risk policies are alike and therefore simply choose the lowest-priced policy for a project. Policies may vary in terms of coverage and exclusions, and the same type of policy often varies between different insurance companies. These variations may be minor, or they may mean the difference between an adequate policy and one that does not meet the requirements of a particular project. Only by reading the policy carefully can one know what is covered, what is excluded, and what coverage may need to be added by means of an endorsement or rider (at an additional premium).

10.1.3 Basic Concepts

Some basic principles apply to all insurance policies, and it is better to understand them to take proper advantage of the coverage to be purchased.

10.1.3.1 Coverage Versus Liability

Coverage refers to the range of risks that the insurance company will pay for should the risk result in an actual loss. There can be liability but no coverage. For example, a party may be responsible for a loss, but not have coverage because it failed to obtain a required policy, or if a policy was obtained, coverage may not be available due to a waiver, lapse, or some other reason. An insured who did not read his policy could not later recover on the grounds of inadequate coverage. It is desired that the insured must thoroughly read and understand any policy which he purchases.

An indemnity is a legal exemption from incurred liabilities or penalties. Many contracts will require that the Contractor indemnify the Owner against any claims. Insurance liability contracts generally provide two benefits to the insured: a duty to indemnify for damages resulting from a claim instituted against him and a duty to defend against the claim itself. The cost of defending a construction claim can be enormous, whether or not the Contractor ultimately is found liable. The duty to defend may exist even where there is no duty to indemnify. However, sometimes an exclusion clause is added by the insurer restricting the defense obligation to only those suits alleging injury or damage covered by the policy.

10.1.3.2 Notice of Loss

All insurance companies require the insured to provide prompt notice in the event of any occurrence which might give rise to a claim. It is required that the insured shall:

(a) Immediately notify the insurer by telephone or telegram as well as in writing, giving an indication as to the nature and extent of loss or damage.
(b) Take all reasonable steps to minimize the extent of loss or damage.
(c) Preserve the parts affected and make them available for inspection by a representative of the insurer.
(d) Furnish all such information and documentary evidence as the insurer may require.
(e) Inform the authorities in case of loss or damage due to theft or burglary.

The purpose of the notice requirement is to permit the insurance company to initiate a timely investigation of a claim. As a result, if timely notice is not provided, the insurance company may deny coverage. It is further desirable that if the notice is to be given to the insurance agent, a copy must be given direct to the insurance company. In the event the insurance agent does not forward the notice to the appropriate insurance company, coverage may be lost.

10.1.3.3 Certificate of Insurance

As discussed in Chap. 7, owners generally require general contractors to produce a certificate as evidence that the required insurance coverage is in effect. This certificate is usually required prior to start of work. Typically, a certificate states that the policies cannot be canceled or altered without a notice to the Owner. The Owner may be added to the policy as an additional insured, which would also be reflected on the certificate. General contractors usually require their subcontractors to procure the same insurance the General Contractor is required to carry. The General Contractor should require its subcontractors to produce valid certificates before the subcontractors begin work.

10.1.3.4 Reservations of Rights

When the insurance company receives a notice of claim or loss, it may choose to defend "under reservation of rights." This means that the insurance company will defend the claim and reserves its right to later deny coverage. The insurance company will invoke this provision if it is unsure whether coverage exists or simply as a precautionary measure. The insurance company will provide written notice of the reservation to the insured.

10.1.3.5 Subrogation

Subrogation is a doctrine according to which insurer, who has made payments to the insured on account of a loss suffered at the hands of a third party, may pursue that third party to recover the amount that has been paid.

The process is achieved by transferring the rights of recovery against the third party from the insured to the insurer. The insurer then stands in the shoes of the insured and exercises all of the rights of the insured against the third party to recover what was paid out against a claim. In other words, the insurer is entitled to assume the legal position of the insured (by suing in the insured's name) to recover from the third party the paid amount.

For example, suppose an Architect retains a HVAC (heating, ventilating, and air-conditioning) specialist on a 20-story apartment building. The HVAC specialist prepares the plans and specifications, and accordingly, the work is done on the project. After completion of the building, when the HVAC system is commissioned, it does not perform as intended. On inspection by experts, it is found that uniform flow is poor from fans, required pressure classification is inadequate, and ductwork is not reinforced to proper pressure classification resulting in duct work separation. The Owner suffers huge losses in rectifying the HVAC defects. The Owner then sues the Architect. The judgment is awarded against the Architect and the insurer defending the Architect pays the claimed amount. Now in place of the Architect, the insurer has the right to sue or subrogate against the HVAC specialist who wrote the specifications.

It may be noted that there is no right of subrogation against anyone who is also insured under the same "loss insurance" policy. That is the reason the standard form of contracts recommends insurance policies to be in joint names of the Owner and the Contractor and sometimes also subcontractors.

Sometimes, a waiver of subrogation clause is also placed in the professional services contract to minimize lawsuits and claims among the parties. In fact, the risk of loss is agreed among the parties through insurance, and the cost of the insurance coverage is contractually allocated among the parties on agreement. The risk, once assigned to the insurer by the parties, is considered to stop at that point due to the absence of a subrogation clause allowing the insurer to seek redress from the party at fault.

10.1.3.6 Indemnification

Indemnification is commonly used in construction contracts to transfer risk inherent in the construction process from one party to another. Most of the standard forms of contract require that the Contractor and Owner both shall indemnify and hold harmless the other, his personnel and respective agents against and from all claims, damages, losses and expenses, suits, or proceedings by third parties, directly or indirectly in respect of bodily injury, sickness, disease or death, or damage to or loss of any property.

Indemnification requirements are further explained under Sub-clause 12.1 of CCDC-2 [2] and under Sub-clause 17.1 of *FIDIC Red Book* [3].

10.1.3.7 Umbrella Liability Coverage

Umbrella liability policy revolves around the same concept as general liability insurance but provides additional coverage limits over the limits provided in general liability insurance. Umbrella or excess liability increases the limits of insurance while providing the same or more restrictive coverage than the primary policies.

When an insured is liable to someone, the insured's primary insurance policies pay up to their limits, and any additional amount is paid by the umbrella policy (up to the limit of the umbrella policy). If the limits on the primary policies are exhausted, or there is no coverage under the primary policy, the umbrella will be considered as a primary policy. The Owner should be named as an additional insured under the Contractor's or Subcontractor's umbrella liability policies.

10.1.4 Contract Requirements

The legal instrument dividing the risk/responsibilities related to natural, technical, and human hazards, between the Owner and the Contractor, is the form of contract. Standard forms of contract usually contain many clauses which oblige the Contractor to insure the works against loss or damage. The Contractor usually has no choice, in the matter as it is standard practice that construction contracts require the contractor to provide certain insurance coverage.

For a proper risk assessment, a thorough study of the contract wording is highly required during the stage of bidding. The study must cover the clauses regarding obligations to insure, special conditions, the drawings and specifications, and the bill of quantities. With regard to contractual insurance requirements, it is always good practice for the Contractor to submit a copy of the Contract Documents to his insurance company at the bidding stage and before starting the works. The Contractor, in general, is not an insurance expert and is not really competent to evaluate the risks and liabilities placed on him by the contract. His insurance company or its agents are qualified to understand the documents and advise him concerning the insurance needs dictated by the language of a given construction contract.

The contract clauses which give rise to the necessity of insurance coverage to safeguard against the risks of loss or damage, stated in most of the contracts, are almost same. Generally, the standard forms of contracts first allocate the various kinds of risks of loss and damage to both the Owner and the Contractor and then impose responsibility for insurance against those risks. Some of the examples of such clauses are as follows.

10.1.4.1 Subcontracting

Usually, the General Contractor is fully responsible for the acts, defaults, and neglects of any Subcontractor, suppliers, his agents, and workers.

Generally, the entirety of the project responsibilities remain with the General Contractor, hence, under such sub-clause, the General Contractor retains responsibility for works or services which he subcontracts.

10.1.4.2 Defective Work

Under this sub-clause, the Contractor is responsible for the overall execution of the works and for the remedying of any defects observed in accordance with the contract provisions.

10.1.4.3 Site Operations and Methods of Constructions

This sub-clause usually states that the Contractor shall be solely responsible for construction means, methods, techniques, and procedures for the various parts of the work under the contract.

10.1.4.4 Temporary Works

Under such sub-clause, the Contractor is required to be solely responsible for the design, erection, operation, and removal of temporary works.

10.1.4.5 Subsurface Conditions

This sub-clause usually states that the site investigation report done by the Owner of the project shall be made available to the Contractor; however, the Contractor shall be responsible for his own interpretation of data on hydrological and subsurface conditions and shall be deemed to have examined and inspected the site and to have satisfied himself before submitting his bid.

This means that all risks related to site conditions mentioned under this type of clause are the Contractor's responsibility. Rain, flood, and climatic conditions can be heavy risks at many construction sites. The Contractor, therefore, must carefully examine the hydrological and geological data relevant to a project site, together with the measures planned for loss prevention and protection of the works.

10.1.4.6 Construction Safety, Security, and Protection of the Environment

This sub-clause provides that the Contractor is solely responsible for the safety of persons on site and for installation of such facilities as will protect the works, the public and the environment from injury or damage.

10.1.4.7 Care of Works

Per this sub-clause, the Contractor remains liable for any loss or damage to the works, materials, and plant to be used during the construction of the project.

10.1.4.8 Warranty

According to warranty sub-clauses, the Contractor remains responsible for about 1 or 2 years (depending on the contract provision) after completion for the performance of the works completed. The Contractor has to rectify any defect if it occurs during maintenance period.

10.1.4.9 Damage to Persons and Property

This sub-clause requires that both the Owner and the Contractor shall each indemnify and hold harmless the other from and against all claims losses, damages, suits, etc. whether losses suffered by them or by third parties, as a result of the execution of the works.

10.1.4.10 Owner's or Employer's Risks

Per this sub-clause, certain risks which are outside the control of the Contractor, such as war, rebellion, civil war, riots, ionizing radiations, nuclear explosives, etc., remain the responsibility of the Owner. This is discussed in more detail under the Force Majeure section below. As a general rule, such risks are retained by the Owner. Generally the property insurer considers these risks uninsurable.

Another risks allocated to the Owner is that if the Owner occupies some portion of the building, any loss or damage due to that will be the Owner's responsibility. Usually, use and occupancy terminates the construction period coverage under the insurance policies. This is called an operational risk which has to be carried or insured otherwise by the Owner.

Additionally, the Owner is also responsible for loss or damage that may arise due to error in design carried out by Owner/Engineer. The Engineer's design is usually insured by professional liability insurance.

10.1.4.11 Force Majeure

Force majeure literally means "greater force." An event that no human foresight could anticipate or which if anticipated is too strong to be considered. Force majeure refers to events outside the control of the parties. These clauses excuse a party from liability if some unforeseen event beyond the control of that party prevents it from performing its obligations under the contract. Typically, force majeure clauses cover natural disasters or other "acts of God" or the failure of third parties. Force majeure clauses are usually applicable to progress failures caused by:

(a) Natural disasters – earthquakes, volcanic activity, hurricanes, and floods
(b) Wars, riots, nuclear explosives, fires, or other major upheaval
(c) Performance failures of parties outside the control of the contracting party – government restrictions, power failures, strikes or lockout by persons other than the contractor or subcontractors personnel, etc.

It is always helpful if the clause sets forth some specific examples of acts that will excuse performance under this clause. Provisions related to force majeure are addressed by CCDC 2 [2] at Sub-clause 6.5.3 and by FIDIC [3] conditions of contract at clause 19. However, FIDIC Conditions of Contract provides five examples of unexpected circumstances, all of which are events that are outside the control of either party. Inclusion of examples helps to make clear the parties intent that such clauses are not intended to apply to excuse to perform for reasons within the control of the parties.

It is also important to note that force majeure clauses are intended to excuse a party only if the failure to perform could not be avoided by the exercise of due care by that party.

10.1.5 Types of Insurance Coverage

Most of the standard forms of contract require in general following two types of insurance covers.

1. General liability insurance policy
2. Loss insurance or property insurance policy

CCDC 2 [2] Sub-clause 11.1 requires that the Contractor shall provide, maintain, and pay for the following insurance coverages:

1. *General Liability Insurance*: in the name of Contractor, the Owner, and the Consultant but only with respect to liability, other than legal liability resulting due to their negligence or arising out of the operations of the Contractor with regard to the work.
2. *Automobile Liability Insurance*: All vehicles working or entering into a construction site using owned, non-owned, or hired vehicles must carry automobile liability insurance.

3. *Aircraft or Watercraft Liability Insurance*: if used directly or indirectly in the performance of work and should be in the joint names of all parties mentioned below in property insurance.
4. *Broad Form Property Insurance*: in the joint names of the Contractor, Owner, and Consultant including all subcontractors.
5. *Boiler and Machinery Insurance*: in the joint names of the Contractor, Owner, and Consultant including all subcontractors.

As per FIDIC conditions of Contract (Red Book) [3], the contractor is also required to procure and maintain the following insurance policies as required by Sub-clauses 18.2, 18.3, and 18.4:

1. *Insurance for Works and Contractor's Equipment (Sub-clause 18.2)*: in the joint names of the parties (means the Employer and the Contractor). FIDIC also specifies that the insurance should cover works, plant materials and contractor's documents, costs of demolition, removal of debris, and professional fees and profit.
2. *Insurance Against Injury to Persons and Damage to Property (Liability Insurance) (Sub-clause 18.3)*: in the joint names of the parties and shall be extended to cover liability for all loss and damage to the Employer's property except things insured in Sub-clause 18.2.
3. *Insurance for Contractor's Personnel (Liability Insurance)*: the insurance is required for the Contractor's personnel, the Employer, and the Engineer.

For more details including the amount of coverage and time limits, *FIDIC Red Book* [3] and CCDC 2 [2] and CCDC 41 [4] (for amount coverage) can be referenced. Here, we will briefly describe liability and property insurance policies.

10.1.5.1 General Liability Insurance Policy (GLI) (Third-Party Insurance)

Under GLI policy, an insurer undertakes that if the insured person (the Owner) becomes legally liable to someone else, i.e., a third party, the insurer will indemnify the Owner against damages and legal costs which become payable [5]. This means that the insurance company takes over the risk from the insured. Third-party insurance for a car driver is an example of this type of policy.

GLI policy covers claims made by third parties for loss arising out of physical injury or damage to persons or property which may occur out of the Contractor's performance of the contract. As mentioned above, insurance liability contracts generally provide two benefits to the insured: a duty to indemnify for damages resulting from a claim instituted against him and a duty to defend against the claim itself which may include costs and expenses of litigation or legal fees recovered by any claimant from the insured and any other costs of defending a claim.

In order for GLI to apply, the liability event must be caused by an "occurrence" of a covered item during the policy period, not by when it is claimed. The occurrence is usually defined in the policy as an "accident." The term "accident" means a mishap or occurrence which is not expected or designed, including liability events arising out of the insured's negligence.

Exclusions Like a loss insurance policy, this policy must also be read carefully to determine what potential losses are excluded from coverage. It should be noted that construction insurance may be invalidated if the insurers are not made aware of changes in the nature, extent, or program for the execution of the works. As the policy is taken by the Contractor, it is his obligation to notify the insurer of changes. If the Contractor fails to arrange or maintain any of the insurances which he is required to provide under the contract, then the Employer may do so at the cost of the Contractor.

In general, a GLI policy does not cover damage to the insured's own person or property including insured's own products and own work.

10.1.5.2 Property Insurance Policy (Loss Insurance Policy)

Property insurance also called a loss insurance policy usually covers sudden physical loss or damage to works, under construction, from any cause (usually fire, vandalism, theft, collapse, storm, lightning and similar casualty losses) except specified exclusions, such as war, riots, nuclear explosions, forces of nature, faulty design, defective material or workmanship, etc. The coverage includes, not only the work under construction, but materials stock piled and plants on site for use.

Most forms of contract provide that property insurance should be in the joint names of the Owner and the Contractor. This is required to avoid the problems of two parallel policies for Employer and Contractor applying to the same project separately. However, as mentioned above, CCDC 2 [2] recommends obtaining "Broad Form" Property Insurance which should be in the joint names of the Contractor, the Owner, and the Consultant including all subcontractors. To have all the project participants named as insured in the same policy basically eliminates the right of subrogation.

During the maintenance period after completion of the project, the insurance is only against damage which the Contractor is required to repair under the terms of the maintenance or defect liability clauses, and so the Owner has no insurable interest in this part of the policy. This section of the insurance could therefore be in the name of the Contractor alone. Similarly, the Owner has no insurable interest in the Contractor's equipment; it could also be insured in the name of the Contractor alone.

Loss due to faulty design and defective material or workmanship are usually excluded by the insurer, therefore costs incurred to rectify for reasons that affect the quality of the project are losses that must be covered by a performance bond or other means. Loss policies also contain exclusions for any type of consequential or indirect loss.

All insured must determine the type of insurance required by the contract and carefully read the policies to determine if they provide the type of coverage that is required, or desired, for that project. Add-on or additional coverage is always available with insurance companies if any party needs to extend the coverage.

10.1.5.3 Contractor All Risk (Car) Policy

The most desirable form of Contractor's insurance is the "all risk" policy form. It is designed to cover mostly all types of engineering projects. A CAR policy provides the parties to the contract with comprehensive and adequate protection against physical loss or damage in respect of the contract works, construction plant and equipment, as well as for third-party claims in respect of property damage or bodily injury arising in connection with the execution of a project. Contractors' all risk (CAR) insurance bridges all risks into a common policy and helps cover the gap between exclusions that would otherwise exist when using separate policies.

CAR insurance is typically taken out jointly by both the Contractor and the Owner, with other parties such as financing companies having the option of being named to the policy. Because multiple parties are included in the policy, they each retain the right to file a claim against the insurer, although all parties also have the duty of informing the insurer of any injuries and damages that may result in a claim.

The coverage begins the commencement of work or after unloading of first consignment at the project site, whichever is earlier and terminates on handing over of works to the Owner or expiry of policy, whichever is earlier. Engineering contracts contains certain conditions regarding responsibilities of the Contractor for effecting insurance coverage. Although these conditions vary according to the country where the works are to be carried out, the policy is being designed to cover the risks normally specified in the contract.

The primary intention of this insurance coverage is to offer a policy, which should respond to most insurance needs on a construction site. It is the only one policy, which covers all interested parties, including the subcontractors and suppliers regardless of the type of damage to the property or who caused the damage. Although the CAR policy is designed to provide broad coverage, it does not cover every risk related to the construction project. Risks usually covered include fire, flood, wind, earthquakes, water damage and mold, construction faults, and negligence. They typically do not cover normal wear and tear, willful negligence, or poor workmanship. The policy can be extended to cover additional items. The insurance coverage is usually provided subject to certain exclusions, conditions, provisions, and payment of the premium agreed between the parties.

A CAR policy usually contains two sections:

Section I-Material Damage Covering physical loss, damage, or destruction of property insured by any cause, other than those specifically excluded in the policy.

Section II-Third-Party Liability Covering the legal liability falling on the insured Contractor as a result of bodily injury or property damage belonging to a third party.

CAR policy also contains two types of exclusions: *general exclusions* and *special exclusions* to each section [6]. For complete details with respect to CAR policy, related reference provided within "reference and further reading" section can be referenced.

10.1.5.4 Professional Indemnity Insurance

Professional liability insurance is also known as error and omission insurance. Professional consultants are normally required by the owners to obtain a professional liability policy to indemnify them against liability for professional negligence. This protects the professional from contractual breaches of the Consultant agreement arising out of negligent acts, errors, or omissions during performance of their services.

The primary reason for professional liability coverage is that a typical general liability insurance policy will only respond to bodily injury, property damage, personal injury, or advertising injury claim. But various professional services and products can give rise to legal claims without causing any of the specific types of harm covered by such policies. Common claims that professional liability insurance covers are negligence, misrepresentation, violation of good faith and fair dealing, and inaccurate advice. Coverage does not include intentional wrongdoing, breach of warranty, criminal prosecution, nor all forms of legal liability under civil law, only those specifically enumerated in the policy.

The standard of care applicable in most instances is whether the professional failed to exercise the required services ordinarily and reasonably well. The Consultant is not held to a standard of perfection and is not required to guarantee error-free plans or a project without any defects. The Consultant may produce plans containing errors and yet be found to have exercised reasonable diligence and therefore not be liable for the errors. However, with the ever-increasing litigious nature of society, it is prudent for professionals to protect their reputation and assets against claims made by owners for negligent acts, errors, or omissions in connection with the professional services they provide.

In the event the Consultant breaches the applicable standard of care, the insurance policy typically covers all amounts in excess of the deductible up to the limit of liability under the policy. The coverage clause of the policy defines the risks that are insured along with a list of exclusions.

The other important factor is that there are two types of policies, one is known as "occurrence"-based policy and the other is called "claims-made" policy. Occurrence-based policies cover claims in which the insured event occurred during the policy periods, e.g., damage occurring to a part of the building during construction period. General liability insurance is written on an occurrence basis.

In contrast, professional liability insurance policies are generally set up on a claims-made basis. This means that the professional person is covered by claims made during the period of insurance, regardless of when the negligent act took place. For example, if a retaining wall for a facility is completed based on an Engineer or Architect's design and specifications, but over time (say 5 years after construction), huge settlement occurs at the retaining wall resulting in severe cracks throughout the wall structure. The Owner could allege that the fault occurred as a result of the Engineer's error in providing his professional services. The Engineer could be held liable for any rework required to make the retaining wall right. The Owner may sue the professional.

Thus, if the claim is made and the professional no longer maintains errors and omissions coverage for whatever reason, or the policy limit has been exhausted by a series of smaller claims, there is no coverage (defense or indemnity) for the Owner to rely upon.

Hence, the professional must ensure that he is properly covered both at the time of the occurrence and when the claim is made. As there can often be a long delay between an occurrence and a subsequent claim, the professional may need to arrange "runoff" cover for a period of time afterward. "Runoff" insurance is a form of insurance which can be bought to provide coverage for claims arising after the professional has ceased work or retired. As this is a specialist area of insurance, the design professional must seek advice from a suitable experienced insurance, expert, or related lawyer.

One way to minimize such claims is to make sure projects are well documented. The consultants must ensure that they set out specific responsibilities in their contracts with clients beforehand and deal with complaints promptly.

10.2 Construction Bonds

10.2.1 Introduction

A construction bond is a written agreement under which the surety (the guarantor) guarantees that the Contractor (the principal) will fulfill its obligations to the Owner. The concept is that if the Contractor fails to fulfill its obligations under the contract, then the surety will fulfill those obligations or compensate the project Owner for the financial loss incurred.

Construction bonding is a risk management tool used to protect the construction owners. The protection is provided against an adverse event that may cause disruption, failure to complete the project due to insolvency of the Contractor or the project's failure to meet contract requirements. It is important to note that bid bond, performance bond, and payment bonds are not intended to protect the contractors who provide them. Instead, these bonds are intended to protect the Owner of the construction project against Contractor failure and to protect certain laborers, material suppliers, and subcontractors against nonpayment.

Many things can go wrong in a large construction project. Because of this, construction bonds are almost a mandatory prerequisite for most of the public and private projects, and accordingly, standard forms are developed and being used in the construction industry for many years. As per the Surety Information Office (SIO) USA [7], out of 1,424,124 contractors in business in 2007, only 969,937 were still in business in 2009, i.e., a failure rate in 2 years was 31.9%. Hence, by specifying surety bonds, the Owner has peace of mind that a sound risk transfer mechanism is in place.

Like an insurance company, the surety looks like an insurer but a bond is not insurance. There is lot of difference between a bond and insurance; the Table 10.1 clarifies some of the differences:

For contractors, there are many different options available for bonding. However, owners need to read the language of the bond carefully before agreeing to use a certain type of bond for their construction project. Not all surety bonds are the same. Some types of surety bonds offer only limited protection, while others may impose strict limitations on when and how an Owner may file a claim against the bond. When evaluating the type of surety bond to be used, construction owners should always evaluate three key provisions of the surety bond: (1) notice of claims, (2) time limit on claims, and (3) surety takeover obligations [8].

In the event of Contractor failure, the major requirement of a surety is that the Owner must formally declare that the Contractor is in default of its obligations guaranteed by the bond which includes following the proper claim procedure as outlined in the bond. Since the surety is not a party to the contract, it cannot declare a default itself and issue a default termination if the Owner takes no action on Contractor's nonperformance. In order to enforce the bond, the Owner must keep the original bond [9]. A photocopy or facsimile is not valid. Additionally, the bond must be signed by all parties including the surety and then delivered to the obligee (Owner). However, if due to certain reasons the obligee does not receive the signed bond, the obligee has no right to claim on the bond [10].

10.2.2 Types of Bonds

Generally, there are three types of bonds used in construction projects, bid bond, performance bond, and labor and material payment bond; however, Lien bond and maintenance bond are also addressed here briefly.

Table 10.1 Difference between insurance and surety bonds

Insurance	Surety bonds
Two-party agreement	Three-party agreement
Premium reflects losses expected	No losses expected, fee is for surety's guarantee
No right of recovery by insurer (risk transfer)	Surety has legal right to recover loss payment
Ideally, losses out of control of insured	Performance may be in control of contractor
Insurance coverage is triggered by accidental events only	Bond is triggered on the default of the contractor

10.2.2.1 Bid Bond

Most bid-calling authorities require that the bid be submitted along with a bid bond either in the amount specified in the bid documents or 10% of the bid amount. If the bid bond is not posted by the bidder, the bid is rejected on the basis of a non-compliant bid. A bid bond backed by a surety company guarantees the Owner that the Contractor will honor his bid and will sign all Contract Documents if awarded the contract. It provides assurance to the Owner that the bidder has the financial capacity to accept the job for the price quoted in the submitted bid. This is because bond-issuing companies perform comprehensive credit and financial reviews before agreeing to provide bonds for a company. Some bid-calling authorities also accept letters of credit or cash instead of surety.

As explained earlier in Chap. 8, Contract "A" is created once the bidder submits a bid. Accordingly, if the bid documents call for irrevocable bids, a bidder has no right to withdraw the bid bond or change its bid during the specified acceptance period.

Additionally, if bid is accepted by the Owner, the bidder must enter into a construction contract, i.e., Contract "B." Hence, the successful bidder will provide the required performance and payment bonds and also will perform the contract in accordance with the Contract Documents. If the successful bidder fails to honor these commitments, the Owner is protected and can call upon the surety to compensate for any loss up to the amount of the bid bond, usually for the difference between the low bid and the next higher responsive bid.

10.2.2.2 Performance Bond

Like a bid bond, a performance bond is also a written agreement between the Contractor and the surety. The Surety guarantees the Owner that the Contractor will complete the contract in accordance to the terms and conditions of the contract; in the event the Contractor defaults or fails to perform, the surety will either complete the contract in accordance with its terms or provide sufficient funds up to the maximum amount of the bond for such completion.

A performance bond is usually issued by a bank or an insurance company, both of which act as a "surety." A performance bond will protect the Owner against possible losses if a Contractor fails to perform or is unable to deliver the project as per the contract provisions. Sometimes, the Contractor defaults or declares himself into bankruptcy. In those situations, the surety is responsible to compensate the Owner for losses. Such compensation is defined as the amount covered under the performance bond.

If the Contractor defaults or is terminated for default by the Owner, the Owner may call upon the surety to complete the contract. A performance bond by CCDC −221 has a condition that any suit or action must be commenced before the expiry of 2 years from the earlier of the date of substantial performance of the contract or the date on which the Contractor is declared in default by the Owner [11]. It may be noted that all bond forms for surety's obligation are always triggered by the Contractor's default; however, the terms "breach" and "default" are not the same in construction suretyship law. The difference in two terms was further clarified by United States Court of Appeals, Fifth Circuit (1994) in a case of L&A Contracting Company v. Southern Contracting Company Inc. as follows [12]:

> Although the terms "breach" and "default" are sometimes used interchangeably, their meanings are distinct in construction suretyship law. Not every breach of a construction contract constitutes a default sufficient to require the surety to step in and remedy it. To constitute a legal default, there must be a (1) material breach or series of material breaches (2) of such magnitude that the obligee is justified in terminating the contract. Usually the principal is unable to complete the project, leaving termination of the contract the obligee's only option.

FIDIC [3] standard form of performance bond has referred to a list of default events against Sub-clause 15.2 "Termination by Employer" of the Conditions of the Contract. These events are explained in Chap. 16 in detail and are briefly stated as under:

1. Fails to provide performance security within time and amount specified
2. Abandons the work or have intention not to continue
3. Subcontracts whole of the work or assigns the contract without the required agreement
4. Becomes bankrupt or insolvent
5. Gives or offers bribe
6. Fails to proceed with the requirements for commencement of the work, time for completion, and submission of program of work (in accordance with clause 8)
7. Fails to comply with a notice for rejection and rectification of work as per clause 7.5.

Many performance bonds give the surety three or four choices, such as completing the contract by the original Contractor with the consent of the Owner or selecting a new Contractor to contract directly with the Owner or allowing the Owner to complete the work with the surety paying the costs or takeover and complete the contract itself.

Owners also have obligations under the performance bond which, if they fail to comply with, the surety is entitled to deny liability under the performance bond. The most common conditions under the performance bond are:

1. Timely notice to the Contractor and surety for considering default.
2. The Owner himself is not in default; for example, if the Owner is in default of its payment obligations, the Contractor's refusal to continue construction may be justified and the surety may refuse to honor.

Hence, it is important that the Owner must read the performance bond carefully and adhere to the provisions of bond before any action is taken under the performance bond.

10.2.2.3 Labor and Material Payment Bond

It is not uncommon in the construction industry for an Owner to require a General Contractor to provide a labor and material payment bond. Protection is provided to all subcontractors and suppliers on a construction project under a labor and material payment bond. Like other bonds, it is also a written agreement between the Contractor and the surety. However, under this form of bond, the surety guarantees the Owner that subcontractors and suppliers will be paid the monies that they are due from the Contractor should the Contractor default in his payment obligations to them.

Under this bond, only a Subcontractor or supplier on the bonded contract can put forward a claim, not the Owner. Regardless if a Contractor is in default of his contract with the Owner or not, a Subcontractor or supplier is able to register a claim. The Owner benefits indirectly from a payment bond in that the subcontractors and suppliers are assured of payment and will continue performance. The amount in a payment bond is often less than the total amount of the underlying contract and is intended to cover anticipated Subcontractor and supplier costs. One of the added benefits of the labor and material bond is that they assist in keeping a project lien free. The bond provides a convenient alternative to liens; however, it is possible that a claimant may file a lien and also claim under the bond.

Since the General Contractor will be the principal on the bond and will enter into it for the benefit of the Owner, only he usually receives a copy of the bond. Subcontractors and suppliers, who would be the claimants under the bond, are generally not provided with a copy of the labor and material bond for a project unless they ask for it. That does not mean that they are not entitled to see the bond. Ontario Construction Lien Act [13], section 39, provides rights to sub-trades to have access to information and to see a copy of the labor and material payment bond. It is therefore important that subcontractors or suppliers request the labor and material payment bond when they enter into an agreement with a General Contractor.

The bond should be reviewed when it is received. The payment bonds set out the method and conditions for making a claim under the bond. Under the bond, anyone who has direct contract with the General Contractor for the provision of labor or material is entitled to put forward a claim. For a claim, the Subcontractor or supplier has to submit a timely notice in writing to the General Contractor with a copy of the contract or purchase order as a proof. Other requirements of surety must be complied with to support the claim.

10.2.2.4 Lien Bond

A lien could be placed on property by anyone, who has done work or furnished materials during construction of a project. In order to clear the title of the property, a release of lien bond is necessary. Most of the construction contracts have provision that the General Contractor remove any liens that are registered against title to the Owner's property. This is achieved by having a construction lien bond with a court. The bond amounts are set by a judge and are based on the amount of a lien. Once the lien bond is arranged with the court, the lien can be vacated from the title to the Owner's property.

A lien bond provides security to a court that a Contractor will honor claims made by its suppliers and subcontractors for payments owing to them for work done on a project in the event a court determines that such claims are valid. The construction Lien Act is further discussed in Chap. 1.

10.2.2.5 Maintenance Bond

A maintenance bond guarantees the Owner of the project that the Contractor will comply with the warranty (maintenance) provisions of the contract. This bond is not very common as a performance bond guarantees all the provisions of the contract and remains in effect up to the end of the maintenance period; however, for some mega projects, this bond is required by the Owner.

Defects and faults in materials, workmanship, and design are all covered under the bond for a specified maintenance period within the contract. Under the bond, the Contractor is obligated to correct any defects that arise, and if the Contractor is unable to solve the defects during the maintenance period or is not in business during the specified maintenance period, the obligee can claim under the bond and is compensated for the defects.

References and Further Reading

1. Bunni, N. G. (1986). FIDIC's view of design liability- construction, Insurance & law. A discussion paper. Published by FIDIC.
2. CCDC-2 (Canadian Construction Document Committee- 2008). Stipulated price contract.
3. The FIDIC. (1999). *Conditions of contract for construction – (Red book).* Author FIDIC.
4. CCDC 41. (2008). *Insurance requirements.* Published by Canadian Construction Documents Committee.
5. Murdoch, J., Hughes, W. (2003). *Construction contracts – Law and management.* Spon Press, London, UK.
6. Jubilee Insurance (Mauritius) Ltd- www.jubileeinsurance.com
7. Surety Information Office (SIO) Washington DC- An article "10 Things You Should Know About Surety Bonds".
8. Davis, S. W. Bricker & Eckler LLP. Article on "*Construction surety bonds: Buyers beware*". Lexology- Globe Business Publishing Ltd.
9. Samuels, B. M., & Sanders, D. R. (2011). *Practical law of architecture, engineering and geoscience.* Toronto: Pearson Prentice Hall, Canada.
10. Glen Boswall, R. Clark Wilson LLP – Article on "Construction bonds guide".
11. Canadian Construction Documents Committee (CCDC)-221. Standard surety performance bond.

12. United States Court of Appeals, Fifth Circuit. (1994). L&A Contracting Company v. Southern Concrete Services.
13. Ontario Construction Lien Act R.S.O, 1990, Chapter C30- Last Amended 2010,c.16,sched 2,s.2- Consolidation Period: From July 1, 2011 to e-Laws currency Date 26 Nov 2013.

Chapter 11
Construction Time

11.1 Introduction

Construction contracting can be extremely time sensitive. Most standard forms of contract impose restrictions on the parties as to when certain actions or activities are to be carried out. Some of the construction contracts add a "time is of the essence" clause which means the deadlines specified in this agreement are vital and mandatory to the contract, and any such breach of these conditions is subject to termination.

However, some standard forms of construction contracts do not have "time is of the essence" clause but clearly specify the commencement date (s), completion date (s), and dates for issuing certain notices. The normal rule in construction is that time is not of the essence because in ordinary circumstances there are provisions for delay, such as extensions of time and payment of damages. More commonly in construction, only an action for damages can be pursued and not a contract termination. As a result, when the works run past the fixed completion date, extra costs are sustained by the responsible party through damages.

Because of the importance of time, the Contractor should not just accept the construction period stipulated in the bidding documents but must make his own assessment of the construction time that will actually be required. In this chapter we will discuss time-related requirements and then how to control time.

11.2 Commencement Date

Commencement date means the date upon which the Contractor receives the notice to commence by the Consultant. The commencement date is the start of the completion period.

A. Surahyo, *Understanding Construction Contracts*,
DOI 10.1007/978-3-319-66685-3_11

The appendix to tender usually specifies a time period after the date of the letter of acceptance within which the Consultant has to send the Contractor a notice to commence. CCDC-2 [1] specifies commencement date under the form of agreement between Owner and Contractor. The time limit for sending notice to commence is important as the Contractor may have prior commitments to the subcontractors, suppliers, or others who may request price increases if orders cannot be placed by the Contractor promptly and as promised.

FIDIC Conditions of Contract [2] under Sub-clause 8.1 provides that the commencement date shall be fixed by the Engineer based on the following requirements:

- The Engineer shall give the Contractor not less than 7 days' notice of the commencement date.
- The commencement date shall be within 42 days after the Contractor receives the letter of acceptance, unless a different date is specified in particular conditions.

Moreover, before such notice is sent, the Owner must be in possession of the site and give the possession to the Contractor on the commencement date. Delay in handing over the possession of site, entitles the Contractor for an extension of time and any extra cost if involved.

The Contractor should commence the works as soon as reasonably possible after the date upon which he received notice to commence. If the Contractor has received only some portion of the site on commencement, then as the works proceeds, the Owner should arrange to give the Contractor possession of such additional portions of the site as may be required to enable the Contractor to continue with the construction of the works without delay in accordance with the schedule of works submitted by the Contractor.

11.3 Completion Date

The original contract date for completion is the date set in the contract at which the work is required to be completed. This is usually calculated from the commencement date specified in the Contract Documents. Many contracts simply establish the number of days (calendar days) instead of a specific date of completion. The stipulated date of completion or time period is considered to be essential to the contract and is an important part of the Contractor's obligation.

In Canada, the project closeout procedure passes through two stages representing two important dates: date of substantial performance of the work and "total performance/completion" date. The primary purpose of this procedure is to expedite and simplify the takeover of a project by its Owner from the Contractor who manages its construction. On completion of the first stage of substantial performance of the work, the Owner can use the facility as intended instead of waiting for the complete wrap up or total completion of the project which usually saves time and money.

The Definitions Section of CCDC-2 [1] defines Contract Time, as "the Work from commencement of the Work to Substantial Performance of the Work." The Contract Time is stipulated in the agreement between the Owner and the Contractor.

FIDIC Conditions of Contract [2] under Sub-clause 8.2 requires that for the purpose of taking over by the Employer, the whole work and each section (if any) shall be completed by the Contractor within the stipulated time of completion. However, under Sub-clause 10.2, FIDIC specifies that the Employer can take over for his use part of the permanent works. The Employer can advise the Engineer to issue a taking over certificate for that part, and the Owner would be then responsible for that part of the work.

11.3.1 Substantial Performance of the Work

Section 2 (Chapter 30) of Ontario Construction Lien Act (CLA) [3] provides that a contract is substantially performed:

(a) When the improvement to be made under that contract or a substantial part thereof is ready for use or is being used for the purposes intended
(b) When the improvement to be made under that contract is capable of completion or where there is a known defect, correction, at a cost of not more than:

(i) 3 % of the first $500,000 of the contract price
(ii) 2 % of the next $500,000 of the contract price
(iii) 1 % of the balance of the contract price

In addition to the above provision, CLA further specifies that in case where some part of the improvement cannot be completed for reasons beyond the control of the Contractor or where both parties agree not to complete the improvement, the price of the services or materials remaining to complete the work shall be deducted from the contract price in determining substantial performance.

11.3.2 Total Performance

Regarding total completion of the project, the CLA [3] provides that a contract shall be deemed to be completed when the cost of completion or correcting any known defect is not more than the lesser of 1% of the contract price or $1000.

No milestone in a construction project is more significant than substantial completion because it determines the date that a Contractor is no longer fully liable for delayed completion or for minor incomplete details. The Owner can no longer assess liquidated damages for late completion. The Owner cannot withhold large sums from the contract price; only the costs of repairing or completing deficiency list items may be retained. Substantial completion is typically also the start date for the warranty period.

11.4 Extension of Time

Most of the standard forms of contract contain certain provisions under which the stipulated date of completion, on which the Contractor is liable to complete the work, can be extended. FIDIC Conditions of Contract [2] under Sub-clause 8.4 provides that the Contractor can be granted the extension of time if the work is delayed by a substantial change in the quantity of an item of work, due to exceptionally adverse climatic conditions, due to unforeseeable shortage of personnel or goods or any delay caused by the actions of the Owner or anyone employed by him. However, as per FIDIC [2], the contract requires the contractor to notify the client with a copy to the Engineer in writing, within 28 days after such event which gives rise to an entitlement to an extension. This notice should specify the reasons for delay and identify the relevant event (s).

CCDC-2 [1] under Sub-clause 6.5.1 and 6.5.2 provides that the Contractor not only shall be granted the extension of time but also shall be compensated for reasonable costs due to such delay, if the work is delayed by the actions of Owner or anyone employed by him or due to a stop work order by court or other public authority (at no fault of contractor). However, in subsection 6.5.3, it further provides that Contractor shall be granted an extension of time without compensation of costs if the work is delayed due to labor disputes, strikes, fire, abnormally adverse weather conditions or any cause beyond the Contractor's control, etc. The notice period for extension of time under CCDC-2 [1] (Sub-clause 6.5.4) is restricted a maximum of 10 days only.

In the event the contract time is extended for valid reasons, the Contractor will not be liable to pay damages for delay. Upon receipt of the Contractor's letter and particulars containing reasons for delay, the Consultant has to decide whether those reasons are justified or not. If in the opinion of the Consultant the reasons submitted by the Contractor are not reasonable and justified, the Consultant can reject the request for an extension of time of the Contractor. However, if the reasons quoted by the Contractor are reasonable, the Consultant can accept the request for an extension of time and should identify the new date for completion of the works.

Moreover, on granting an extension of time to the Contractor, the Contractor is required to resubmit the insurance policies and performance bond to cover revised dates.

The common acceptable events mentioned in most of the standard forms of contract entitling the contractor for extension of time are:

(a) Additional works.
(b) Delay in issue of drawings or instructions.
(c) Delay in handing over possession of the site to the contractor.
(d) Exceptionally adverse weather conditions.
(e) Differing site conditions.
(f) Delay on the part of nominated subcontractors/suppliers. Nominated or selected subcontractors are selected by the Owner, and the Owner negotiates the terms of the subcontract with the subcontractors and instructs the General Contractor

to employ that particular Subcontractor. The immediate objective of adopting nominated/selected subcontracts is that the selected parts of the construction works will be carried out by specialist subcontractors selected by the Owner and not by the Contractor.

(g) Public authorities intervention.

(h) Force majeure, the items covered under force majeure are described under FIDIC [2] clause 19, such as riots, strike, explosions, earthquake, hurricane, etc.

11.5 Liquidated Damages

If the Contractor fails to complete the works within the contract completion time, the Contractor will be liable for liquidated damages. Liquidated damages can be defined as the damages that result from the project's extended duration beyond the stipulated date of completion. Usually delay is considered when the activities controlling critical path are delayed which affects the overall project completion.

Liquidated damages are a specific sum of money usually stated in the Contract Documents to be recovered from the Contractor for each day of delay in completion of the contract, subject to the certain limit stated. Liquidated damages are a widely used method of ensuring contractors complete the work in a timely fashion. Without a fixed completion date, liquidated damages will not be assessed. That's why construction contracts usually include a fixed completion date which may be extended through time extension provisions in the contract. The Contractor will not be liable to pay delay damages if the extension of time is approved by the Owner.

Delay damages provisions are regularly used in construction contracts; however, it is important that the rate of liquidated damages must be a genuine pre-estimate to compensate for possible damages and not be so large as to be construed as a penalty. The courts have enforced liquidated damage provisions in construction contracts only when they reasonably represented the actual damages suffered by the Owner. If the amount was excessive and unreasonable, the courts have ruled that such payment by the Contractor to the Owner constituted a penalty and was not enforced.

The amount of liquidated damages is determined by the Owner before bids are invited; bidders are therefore aware of the nature of their commitment at the time of bidding. It is important to specify a limit on the total liquidated damages which may become due. Usually damages for non-completion are indicated as a lump sum amount per day for each and every calendar day delay from the specific time for completion of the works until actual completion of the works.

However, FIDIC Conditions of Contract [2] under Sub-clause 8.7 provides that the payment or deduction of these damages shall not relieve the Contractor from his obligation to complete the works or from any other duties, obligations, and responsibilities under the contract.

11.6 Controlling Construction Time

The control of time, along with cost and quality are the basic goals for project management. Time control is important for both Owner and Contractor, because time savings can improve profits and time overrun causes delay in completion of the project and increases overheads, i.e., losses. Basically control of completion time requires controlling the baseline/planned schedule. The schedule provides the parties with the planned start and completion dates of activities so as to control the works to complete within the specified time. Most of the forms of contract require the Contractor to submit a schedule for execution of works. The contract schedule is usually one of the most important provisions of the contract to all parties. Adequate schedule of works mainly comprises:

(a) The activities
(b) The start and completion date of each activity
(c) The resources required
(d) The construction sequence
(e) The critical path and so on

Adequate scheduling can facilitate the timely procurement of required materials and insure the completion of a project in a reasonable time frame. Whereas inadequate scheduling can result in considerable waste as laborers and equipment wait for the availability of needed resources or the completion of preceding tasks. Delay in the completion of a project due to inadequate scheduling can also create problems for an Owner who has made plans to start using the constructed facility.

The schedule of works allows the Consultant to monitor the progress and accordingly make arrangements for the supply of drawings, issuance of instructions, and coordination with other contractors engaged in the project. It also helps the Consultant to ascertain how the phasing of operations effect any other interrelated contract. For the Contractor, a schedule helps with planning and timing the manpower and equipment required, as well as arranging delivery of materials to meet the planned schedule dates. In the event, the Contractor finds any activity behind planned schedule; he can apply extra efforts to overcome that delay, so as to bring the overall schedule back on time. For all of these reasons, a quality schedule is highly important to all parties in the construction process.

Schedule control requires the establishment of a schedule baseline which is the final schedule of works and can only be changed and updated time to time during the course of construction by formally approved changes. The baseline schedule serves as a reference document for progress control, and meeting the baseline is one of the measures of project success.

The actual work performed is usually compared to the planned work to determine if construction is proceeding satisfactorily. The Project Manager or Contract Administrator must measure how the project is performing by using performance measurement techniques, such as schedule variance and schedule performance index, and be able to recommend and implement corrective and preventive actions to adjust the project during construction to make sure the baseline is maintained. After completion of construction, similar comparison between the planned schedule and actual accomplishments may be performed to allocate the liability for project delays due to changes requested by the Owner, inclement weather, or other unforeseen circumstances.

11.6.1 Scheduling Methods

In general, the type of schedule is independent of the size of the project. Over the past 50 years, several techniques have been developed, and they can be grouped in four major categories:

(a) Charts – bar charts and milestone charts.
(b) Network scheduling – there are several network scheduling techniques; however, two methods, arrow diagrams and precedence diagrams, are mostly used as indicated in Fig .11.1.
(c) Linear scheduling method/line of balance method.
(d) Simulation.

As per a study made by constructions professionals [4], the most popular time planning and control technique is Gantt bar chart, which is used by 35% of contractors and 33% consultants. This is closely followed by critical path method (CPM) used by 28% contractors and 34% consultants. Similarly, the most widely used software is Microsoft Project and is used by 35% contractors and 57% consultants, whereas Primavera is used by 15% contractors and 19% consultants.

Thus, the most commonly used scheduling methods in the construction industry, Gantt bar chart and CPM including software Microsoft project, are briefly discussed here. For a complete study of scheduling procedures, books on construction management or scheduling techniques can be referenced.

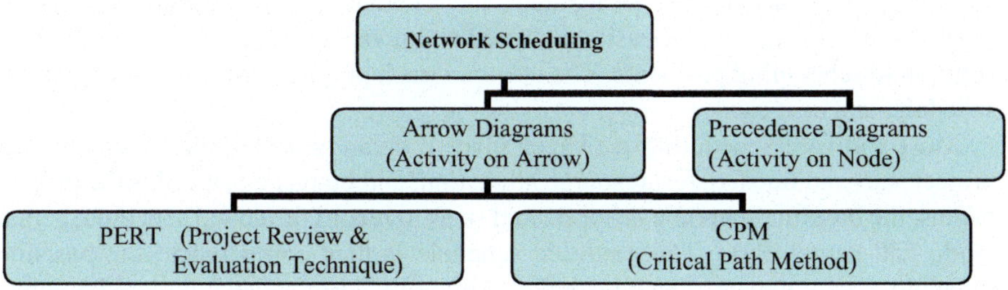

Fig. 11.1 Network scheduling types

11.6.2 Bar Chart or Gantt Bar Chart

This is the most common type of graphical display representing activities/tasks across a time scale. It is named a Gantt chart in honor of the inventor, Henry Gantt, who first utilized this procedure in the early 1900s. This technique is widely used as a scheduling mechanism because it is the simplest and least complex means of portraying progress. It is very easy to understand and a very useful format in which to display the results of a schedule prepared by a more contemporary technique.

A bar chart is a collection of activities in a vertical column with time represented on a horizontal x-axis. The first step is to determine which activities will be listed. To do this, the work has to be broken down into smaller, finite activities, called a "work breakdown structure (WBS)." The next step is to estimate the duration of each activity. Finally the sequence is determined and the bars are plotted on the schedule.

The start and finish dates are shown for each activity, and the duration is indicated by a horizontal bar to the right of the description of the activity. It is the length of the bar, which represents the duration of the activity. The planned duration of each activity can also be calculated and indicated in a vertical column. The unit of time is normally days, weeks, or months. For monitoring progress, the space within the bars is used. Figure 11.2 shows an illustration. The blue bar indicates planned dates and is updated via the upper red bar as the work progresses.

A major shortcoming in the use of these simple type bar charts is that they do not show the interdependencies between events and activities and do not lead to a high degree of control. Therefore, this method is not useful for projects with a large amount of activities and those with complex relationships and dependencies between activities.

11.6.2.1 Computerized Scheduling Techniques

On complex projects, the use of computerized scheduling techniques has become common. Computerized scheduling uses the coordination of each specific detail of the project to build a project schedule. Revisions are easily made and automatically incorporated as the computer recalculates the schedule. Resources (and the costs of those resources) can also be easily analyzed and revised in an effort to find the best solution to scheduling problems. The user can track progress and control resources, schedule, and cost easily.

Most software scheduling packages involve creating a "baseline" schedule of project start and finish dates, resource allocation, and budget costs. After the project begins, the baseline schedule is duplicated so as to record the actual schedule, actual work, and actual costs. The modified schedule is then compared to the baseline schedule to spot delays, check resource availability, keep a check on budget items, and allow the user to look at alternatives to keep the project on track.

ID	TASK NAME	START DATE	FINISH DATE	DURATION	YEAR 2015		
					January	February	March
◇	**New Store Construction**	Jan-5-15	Mar-27-15	60 Days			
1	Mobilization	Jan-5-15	Jan-9-15	5 days			
2	Foundations	Jan-12-15	Jan-23-15	10 days			
3	Backfill & grade slab	Jan-26-15	Feb-2-15	6 days			
4	Columns, Beams& Roof Slab	Feb-3-15	Feb-12-15	8 days			
5	Masonry & Plaster	Feb-13-15	Feb-23-15	7 days			
6	Doors & Windows	Feb-24-15	Mar-4-15	7 days			
7	Plumbing & Electrical	Mar-5-15	Mar-11-15	5 days			
8	Painting & Tiling	Mar-12-15	Mar-23-15	8 days			
9	Cleaning & Handing over	Mar-24-15	Mar-27-15	4 days			

Note: Bar lines show approximate dates.

Fig. 11.2 Basic project bar chart (completion period 60 calendar days (except Saturdays and Sundays))

There are so many computer software tools which use Gantt chart scheduling techniques, like TurboProject, Milestones Professional, BuildIT Systems, Primavera P6, Microsoft Project, etc. However, as mentioned earlier, we will address Microsoft Project only.

11.6.2.2 Microsoft Project (MS Project)

MS Project is a powerful scheduling tool, and its Gantt chart can be used to create a schedule, as well as to track it and print reports. It also provides a calendar that allows the user to review, create, or edit the project and to see quickly which tasks are scheduled on particular days, weeks, or months.

The MS Gantt chart displays task information as both text and graphics. The Gantt table lists information about each task, and the Gantt bar chart displays task duration, start and finish dates on a timescale. The relative positions of the Gantt bars show which tasks come before or after or overlap the task represented by each bar.

The MS Gantt chart can be used to:

(a) Create a project by entering tasks and task durations.
(b) Establish a sequential relationship between tasks, so users can see how changing task duration affects the start and finish dates of other tasks and the project finish date. Tasks can be moved, split, re-split, and recombined into whatever portions the user requires. The duration of the split task can be changed only when the user changes the duration of a given portion.
(c) Review the tasks that are scheduled on particular days, weeks, or months.
(d) Adjust working time for the entire project or for a specific group of resources. The user can also specify all nonworking days along with official and national holidays.
(e) Create overlapping or delaying relationships between tasks by using lead and lag time. With lead time, the user can schedule an overlap between two tasks, so that one task begins before its predecessor task finishes. With lag time, the user can delay the start of a successor task. The user can express lead or lag time as units of time or as a percentage of the predecessor task duration.
(f) Assign personnel and other resources to tasks so as the user can use those resources as efficiently as possible.
(g) Assign costs to tasks and resources, allowing the user to monitor cash flow of the project.
(h) Establish a baseline schedule (planned schedule).
(i) Track progress by comparing planned and actual start and finish dates and by checking the percentage of each task that is complete. MS Project provides a tracking table which is a very convenient way to view and update the data.
(j) Identify the critical path of a project. The critical path of a project is the sequence of tasks from the start date to the finish date that must finish on time for the entire project. Any delay among tasks on the critical path will delay the finish date of the project. The critical path may change from time to time as activities are completed ahead of or behind schedule.
(k) Print specific project information that meets the needs of a particular person or group, including:

    - Overview information, such as a summary report
    - Task information, such as a Gantt chart
    - Resource information, such as a schedule for each resource
    - Cost information, such as a list of over-budget resources
    - Tracking information, such as a list of tasks that are behind schedule

The latest version available is MS Project 2013. It carries new features like:

- Share meetings.
- Save and share files to your own SkyDrive or to your organization's site.

- Work from almost anywhere, even on PCs that don't have Project 2013 installed.
- Entirely new set of pre-installed reports.
- Burndown reports showing planned work, completed work, and remaining work as lines on a graph providing at-a-glance status of project.

11.6.3 Critical Path Method (CPM)

In 1950s, DuPont Corporation and Remington Rand, in an effort to improve project schedule techniques, developed a scheduling system called the Critical Path method or CPM scheduling. CPM can be defined as a management planning and analysis tool which makes use of a graphic display called a network to depict the essential relationship and sequence between various activities and events required to achieve the end objectives for a particular project.

CPM requires a project to be broken down into component activities that can be developed into a network diagram showing sequential relationships of tasks. The technique uses arrows and lines to represent the relationship of activities to one another and "nodes" or circles to represent the events at the point in time when the tasks should be started or completed. Events and activities are numbered for identification on the network. The duration of a task is estimated, and critical activities are defined that must be kept on schedule if the entire project deadline is to be met. Also built into schedule is "float" or extra time available to secure the timely performance of an activity.

Critical path method's graphic process portrays the logic of the construction plan. The primary advantage to CPM scheduling is the ease with which the technique can handle many work activities on complex projects. Critical path methods are concerned, not only with the sequence and interrelationship of activities but also determine the time and cost of completing activities. CPM provides a standard method of documenting project plans, resource allocation of all phases of the plan, schedules, and time and cost performances.

11.6.3.1 Basic Rules for Drawing Networks

1. Every arrow (activity) must begin and end with a node/event as shown in Fig. 11.3.
2. Junctions of arrows are called "events" or "nodes." They are a point in time and consume no time. They are numbered to provide a convenient numeric sequential designation for all activities. The number at the tail event is always smaller than the number at the head event. In large networks, it is advisable to use multiples of tens, so that any future changes or additions can be easily incorporated.
3. No two nodes in the network can be directly connected by more than one arrow.

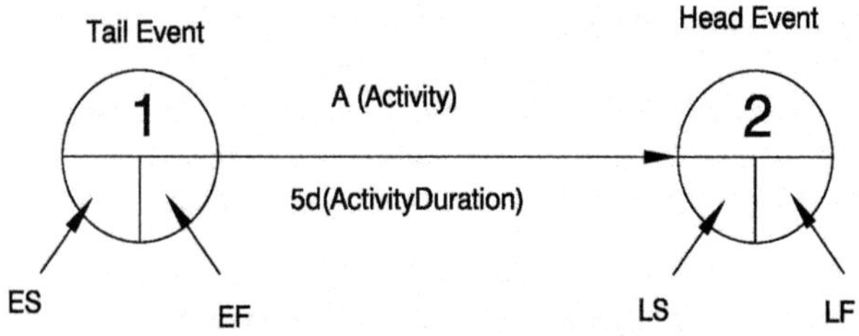

Fig. 11.3 Tail and Head Event Concept (*ES* early start, *LS* latest start, *EF* early finish, *LF* latest finish, *1* and *2* event numbers)

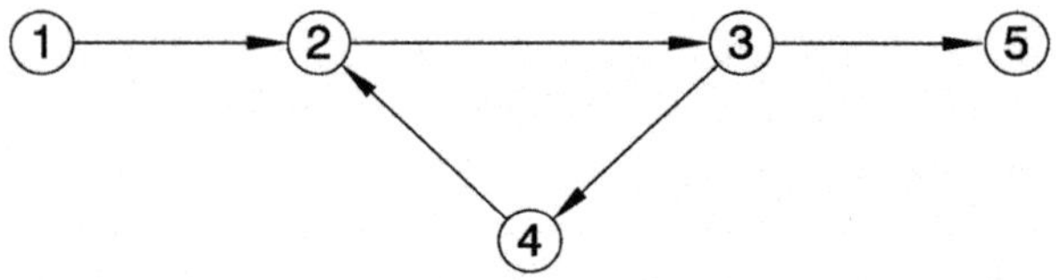

Fig. 11.4 Looping of Activities (Activities *2*, *3*, and *4* makes a loop)

4. Only one arrow is used to represent the operation to be performed. The length of the arrow and the direction in which it points have no significance. A restriction is made that each arrow is unique, and any single activity cannot be represented by more than one arrow. All arrows in the network should be directed, more or less, from left to right.
5. A complete network should have only one point of entry (a start event) and only one point of exit (a finish event).
6. Every activity must have a preceding activity or tail event and a succeeding activity or head event.
7. No activity can start until its tail event has been reached. Consequently, overlapping operations are prohibited. If they occur, they must be broken down into two or more activities representing those portions of the operation to be completed before later portions are commenced. The overlapping of work as is done on a conventional bar chart construction program is impossible with CPM, and hence this new technique offers a much greater degree of control over all the operations on site.
8. An event is not complete until all the activities leading into it are complete.
9. A series of activities leading back to the same event (looping) as shown in Fig. 11.4 are not allowed.
10. To show logic relationship, dummies can be used. Dummies are represented in a network by a broken arrow. They do not consume any resource or time.

11.6.3.2 Steps in CPM Project Planning

(a) Prepare a list of activities.

Prepare a list of the project's activities from the work breakdown structure and assign a code to each activity, as shown in Table 11.1. This list can be used as the basis for adding sequence and duration information in later steps.

Table 11.1 Scheduling of activities

Activity	Code	Arrow	Duration (days)	EST	LST	EFT	LFT	TF	FF	Remarks
Formation cleaning and compaction	A	1–2	5	0	0	5	5	0	0	Critical
Blinding	B	2–3	5	5	5	10	10	0	0	Critical
Formwork erection	C	3–4	10	10	10	20	24	4	0	–
Alignment and leveling formwork	D	4–6	3	20	24	23	27	4	0	–
Cutting and bending steel	F	3–5	10	10	10	20	20	0	0	Critical
Dummy	Dotted	5–6	0	20	20	23	27	7	3	–
Laying polythene sheet	E	6–7	3	23	27	30	30	4	4	–
Fixing steel and spacers	G	5–7	10	20	20	30	30	0	0	Critical
Concreting	H	7–8	3	30	30	33	33	0	0	Critical

(b) Estimate duration of each activity.

The time required to complete each activity can be estimated using past experience or the estimates of a knowledgeable person. CPM is a deterministic model that does not take into account variation in the completion time, so only one number is used for an activity's time estimate.

(c) Determine the sequence of those activities.

Some activities may proceed concurrently, whereas, certain activities are dependent on the completion of others. For each activity, one must ask:

1. What activities precede this activity?
2. What activities follow this activity?
3. What activities run concurrently with this activity?
4. What controls the start?
5. What controls the finish?

In this way, each activity is examined, and the necessary sequences of activities are determined. Every activity has a definite event to mark its possible beginning; this event may be the start of the project itself or the completion of a preceding activity.

(d) Draw a network diagram.
(e) Assign numbers to the events.
(f) Prepare activities chart with activity name, code, events, duration, EST, EFT, LST, LFT, TF, FF, and IF.

The activities chart as per Table 11.1 is prepared based on all above information. This table shows the results of all the time and float calculations for the proposed network. These network times are defined as under:

EST = earliest start time = EFT (minus) – activity duration. The time at which an activity in a project can begin.

LST = latest start time = LFT – activity duration. The time at which an activity can commence, if the minimum project duration is to be maintained.

EFT = earliest finish time = earliest tail time + activity duration.

LFT = latest finish time = latest head time – activity duration.

The EST or EFT for each activity is mostly recorded in the left side of the circle. It may be noted that EFT and LFT become EST and LST, respectively, for the following activity. The ESTs and LSTs are calculated using the normal forward and backward passes.

TF = total float = LFT – EFT or LST – EST and is calculated as latest head time – earliest tail time – activity duration. Total float is the excess of available time over the required project end time, by which the start of a noncritical activity may be delayed.

FF = free float = earliest head time – earliest tail time – activity duration. Free float (term float and slack is used interchangeably) is always equal to or less than total float. If a negative value is produced, it is considered as zero. The FF of an activity is the amount of time by which the start of the activity may be delayed without delaying the start of any succeeding/following activity.

(g) Identify the critical path.

When EST and LST or EFT and LFT are identical, i.e., if TF is zero, the activity is critical, and all of the critical activities form the critical path. The critical path is the longest duration path through the network. There may be more than one critical path among all the project activities, so completion of the entire project could be delayed by delaying activities along any one of the critical paths.

(h) Update the CPM diagram as the project progresses.

As the work progresses, the actual activity completion times will be known, and the network diagram can be updated to include this information. Whenever the activity durations are updated, the network is analyzed to determine whether the critical path and the project completion period have been affected. If the progress is found to be running behind planned time, the network may be amended by using schedule compression technique. The objective of the schedule compression technique is to try to compress the schedule without changing project scope. Schedule compression technique includes:

1. Fast tracking technique: Doing critical path activities in parallel that were originally planned in series.
2. Crashing technique: Making cost and schedule tradeoffs to determine how to compress the schedule for the least incremental cost while maintaining project scope. Crashing may include overtime work, additional tools and plants, manpower, working in double shifts, etc.

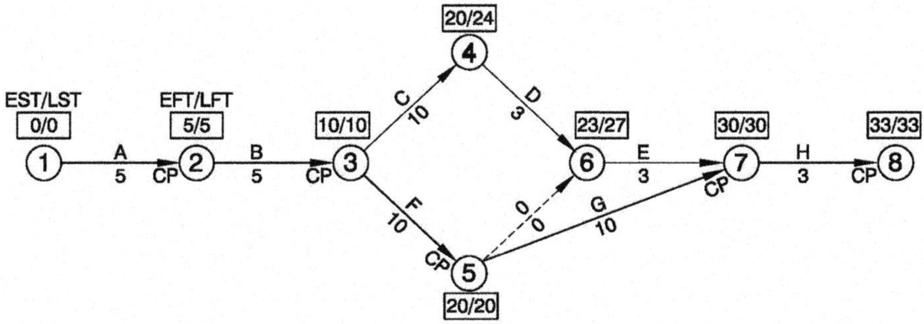

Fig. 11.5 Concreting of foundation for ten houses. Network diagram (*CP* critical path)

In general, the updating is carried as follows:

1. By coloring each arrow to indicate the completed portion of the project
2. By crossing out the original/planned durations of the completed activities and inserted the actual durations in color
3. By crossing out the original earliest and latest finish times in the oval time box of all completed events and inserting the actual completion time in color
4. By indicating the expected completion time of all activities in progress

In this way, the network diagram is being assessed and updated throughout the construction period.

Example
Draw a network diagram for a project "concreting of foundation for ten houses."

In the light of above steps (a) and (b), prepare a Table 11.1 and draw a network diagram (Fig. 11.5) to determine the sequence of activities. Thereafter, based on the duration of activities and with the help of forward and backward passes, calculate times and logical relationships as shown in the table 11.1 and network diagram Fig. 11.5.

11.6.3.3 CPM Software

There are several critical path planning software tools. Most of them carry similar concepts for scheduling required data.

Computer software allows the user to specify tasks and their durations and will then calculate the earliest and latest start and finish dates for each task, based on dependency links which the user inputs. It will calculate *Float*, the amount of spare time that a given task has in the schedule as an indication of how important that task is to the overall timing of the project. It also calculates critical path activities. Most of the CPM software shows all project management data in a spreadsheet pattern. This software includes the function for working time adjustment. They have functions for resource allocation and leveling, cost functions, etc.

References and Further Reading

1. CCDC-2. (2008). Canadian Construction Document Committee – Stipulated price contract.
2. The FIDIC. (1999). *Conditions of contract for construction – (Red book)*. Author FIDIC.
3. Ontario Construction Lien Act R.S.O,1990, Chapter C30- Last Amended 2010,c.16,sched 2,s.2- Consolidation Period: From July 1, 2011 to e-Laws currency Date 26 Nov 2013.
4. Olawale, Y., & Sun, M. (2010). Cost and time control of construction projects: Inhibiting factors and mitigating measures in practice. *Construction Management and Economics, 28*(5), 509–526.

Chapter 12
Construction Payment/Cost

12.1 Introduction

Price and payment are important provisions in every construction contract. Contractors, subcontractors, and suppliers all require a certain amount of cash to finance their ongoing operations. For this reason, payment terms usually are spelled out in great detail in construction contracts. The interrelation between time and money is always of major importance to both the Owner and the General Contractor.

In order to complete a project on schedule, the Contractor must have sufficient cash flow. In the construction industry, funds flow from the Owner to the General Contractor and hence to the subcontractors and suppliers, based on satisfactory work done. This means that each Contractor must be paid promptly, within the terms of the contract for the work performed within the payment period specified. Therefore, it is essential that progress payments, as well as payments for additional work ordered by the Owner, be quickly processed to maintain the Contractor's cash flow. It is also important that claims for additional payment be quickly resolved to avoid a project delay.

The Owner also has to make sure that the payments made against work done and changes made shall not cross the required project budget. For this reason, the Owner must establish a sound cost monitoring and controlling system. This chapter addresses the provisions and issues regarding payments made against work done, changes, holdback, and controlling project cost.

12.2 Contract Price

The contract price is defined in CCDC-2 [1] as the amount stipulated in Article A-4 of the agreement. In FIDIC Conditions of Contract [2], the contract price is considered as the amount accepted in the Letter of Acceptance and any adjustments which

A. Surahyo, *Understanding Construction Contracts*,
DOI 10.1007/978-3-319-66685-3_12

are provided for in the contract. In fact, under FIDIC Conditions of Contract [2], the contract price is distinguished from the accepted contract amount.

The contract price is the sum identified for provision of the entirety of the works. It is not assured that the Contractor will get that amount upon completion. Additions and alterations are part of most contracts which will, of course, change the contract price. There are provisions laid out in most of the contracts on how these changes are to be evaluated in terms of cost.

12.3 Progress Payments

Progress payments to the Contractor are usually made on monthly basis as the work proceeds. CCDC-2 [1] Sub-clause 5.2.1 provides that an application for payment may be made monthly as the work progresses. As discussed in Chap. 2, CCDC-2 [1] Sub-clause 5.3.1.3 further provides that the Owner shall make progress payment to the Contractor on or before 20 calendar days after the Consultant's receipt of the Contractor's application for payment or the last day of the monthly payment period.

Similarly, FIDIC Conditions of Contract [2] also provides under Sub-clause14.3 that the Contractor shall submit the application for interim payment certificate after the end of each month. As discussed in Chap. 2, FIDIC Conditions of Contract [2], clause 14.7(b), further provides that the Employer shall make interim progress payments within 56 days after the Engineer receives the Contractor's statement and supporting documents.

The Contractor submits his interim valuation, based on the current work done and cost of material delivered to site, which is then evaluated and certified by the Engineer. For unit price contracts, the payment is made according to the unit rates for work carried out. For lump sum contracts, interim payments are made on a percentage basis for work done in that month. Whereas, cost-plus contracts usually provide for the Contractor to submit interim payment at specified intervals during the life of the contract. Sometimes mistakes happen in recording measurements and payments are issued based on those incorrect measurements. Thus it is advisable to add a provision in the contract specifying that, "The Contract Administrator may correct any error in an interim certificate in subsequent certificates."

Some contracts specify minimum amounts for interim payment; hence, the Contractor has to submit his interim payment when he crosses or reaches that figure. While evaluating the interim valuation of the Contractor, if the amount due to the Contractor is less than the minimum amount stated in the contract, no certificate is issued and the payment is included in the following certificate. Before certifying the payment, some deductions are made from each interim payment like previous payments, holdback/retention money (usually 10% of the total work done), and any government tax (if applicable). The holdback money is one of the securities held by the Owner to ensure fulfillment by the Contractor of his obligations under the contract.

In order to have the contracting and subcontracting system function smoothly, delay in payment must be avoided. As per CCDC-2 [1], if the Consultant fails to issue a payment certificate or the Owner fails to pay the Contractor due certified amounts within the specified time as mentioned above, then the Contractor may give notice in writing to the Owner that the Owner is in default of his contractual obligations and to advise him to correct the default within 5 working days (Sub-clause 7.2.4); otherwise the Contractor may suspend or terminate the contract.

Similarly, as per FIDIC Conditions of Contract [2], if the Employer fails to issue a payment within the specified time as mentioned above, then Sub-clause 16.1 provides that the Contractor may give 21 days' notice and then suspend or reduce the rate of work. Additionally, under Sub- clause 16.2(c), the Contractor is entitled to terminate the contract if payment is not received within 42 days of the due date.

12.4 Final Payments

Acceptance of the project and final payment by the Owner must proceed in accordance with the terms of the contract.

12.4.1 Procedures Under FIDIC Conditions of Contract

As per FIDIC [2] Sub-clause 10.1, when the Contractor feels the works are completed in accordance with the contract and will be ready to be taken over by the Employer within 14 days, he can give such notice to the Engineer. The Engineer within 28 days shall issue the taking-over certificate to the Contractor, indicating the completion date except for any minor outstanding work or defects, provided that they do not substantially affect the use of the works for their intended purpose.

After receiving the taking-over certificate, the Contractor (within 84 days) shall submit to the Engineer a statement at completion (Sub-clause 14.10) with supporting documents, in respect of the whole of the works, showing in detail the final value of total work done, including further sums which the Contractor considers to be due and an estimate of amounts, which the Contractor considers will become due to him under contract. The Engineer within 28 days of receiving the statement of completion should certify to the Owner the amount of payment to the Contractor, which he considers due and payable at that time.

Thereafter, once the Contractor rectifies the defects within the defects' liability period, the Engineer shall issue the performance certificate (Sub-clause 11.9) within 28 days if he is satisfied that the Contractor has fulfilled his obligations during the defects' notification period. Within 56 days after receiving the performance certificate, the Contractor can submit an application for the final payment certificate (Sub-clause 14.11) with supporting documents, showing in detail the value of total work done in accordance with the contract and any further sums if due. While evaluating, if the Engineer has any comments or needs further information, the Contractor should attend to that requirements and resubmit his final statement/valuation.

However, if there are still differences of opinion which need time to resolve, the Engineer should issue a further interim certificate for the portion of work not in dispute, such that the Contractor can finalize the requirements of that portion. As soon as the dispute is resolved and the Contractor and Engineer agree on the basis of final statement, Sub-clause 14.12 requires that the Contractor should give a written discharge to the Employer with copy to the Engineer, confirming the total settlement of all monies due to him as per contract. Such discharge, however, shall become effective once the final payment has been made and the performance security is returned to the Contractor.

After receiving the final statement and Contractor's written discharge, the Engineer should within 28 days issue a final certificate for payment to the Employer, which in the opinion of the Engineer is due under the contract.

12.4.2 Procedures Under CCDC-2 and Related Canadian Documents

As per clause 5.4 of CCDC-2 [1], when the Contractor has determined that the works are substantially performed (as defined in the construction Lien Act [3]), he can give such notice within 1 working day to the Consultant and Owner including a list of items to be completed or rectified. Upon verifying that the contract has been substantially performed, the Consultant within 20 calendar days shall issue a copy of that certificate to the Owner and the Contractor with the date on which the works were substantially performed. However, if the Consultant identifies some deficiencies in the work, a list of all such deficiencies or uncompleted work will be issued to the Contractor for correction and completion before issuing the certificate of substantial performance.

After receiving a copy of substantial performance certificate, the Contractor shall publish a copy of the certificate in a construction trade newspaper as required under Section 32(1) of the Construction Lien Act [3]. As per Section 31 of the Act, the day following the date of publication shall be the date of commencement of the required 45-day period to release the basic statutory holdback monies. The holdback monies are defined in Chap. 1.

According to the Ontario Provincial Standards (OPS.MUNI 100) [4], when the Consultant issues the certificate of substantial performance, he should also issue the substantial performance payment certificate and the substantial performance statutory holdback release payment certificate which shall include:

1. The value of work performed to the date of substantial performance

2. The value of outstanding or incomplete work
3. The amount of the statutory holdback, allowing for any previous releases of statutory holdback to the Contractor in respect of completed subcontracts and deliveries of preselected equipment
4. The amount of maintenance security required
5. The amount due the Contractor

The payment of the amount certified shall be made within 30 days of the date of issue of the payment certificate except the basic holdback. The basic holdback shall be released after the expiry of the 45-day period as mentioned above but on submission of the following documents as required under Ontario Provincial Standards (OPS.MUNI 100) [4]:

1. Proof of publication of the certificate of substantial performance
2. A certificate of clearance from the Workplace Safety and Insurance Board,
3. A statutory declaration (CCDC- 9A [5]) to state that all liabilities incurred by the Contractor and the Contractor's subcontractors in carrying out the contract have been discharged except for amounts properly retained as a holdback or as identified in any dispute of payments
4. A release by the contractor in a form satisfactory to the Consultant releasing the Owner from all further claims relating to the contract

For final payment, CCDC-2 [1] under clause 5.7 provides that when the Contractor considers that the work is completed, he shall submit an application for final payment. The Consultant, within 10 calendar days shall review the work and, if satisfied that the work is completed in accordance with the contract, shall promptly issue a final certificate for payment except the finishing holdback. As per Ontario Provincial Standards (OPS.MUNI 100) [4], the final payment certificate shall show:

1. Measurement and value of work at completion
2. Invoice for finishing holdback
3. The amount due to the Contractor

Like basic holdback, the finishing holdback shall also be released after the expiry of 45 days from the completion date but on submission of the following documents as required under Ontario Provincial Standards (OPS.MUNI 100) [4]:

1. A written statement by the Contractor (in a form satisfactory to the consultant) releasing the Owner from all further claims relating to the contract
2. A certificate of clearance from the workplace safety and insurance board
3. A statutory declaration CCDC- 9A [5] to state that all liabilities incurred by the Contractor and the Contractor's subcontractors in carrying out the contract have been discharged

Once the Consultant verifies that no liens have been preserved, the finishing holdback is released 1 day after termination of the 45-day period.

The procedures for substantial performance and completion take-over of projects are further explained in more detail in document Number-100 published by Ontario

Association of Architects (OAA) and Ontario General Contractors Association (OGCA) [6].

12.5 Cash Allowances and Contingencies

12.5.1 Cash Allowances

Cash allowances also known as provisional sums for materials or work are added in the schedule of prices or bill of quantities when it is not possible to clearly define the scope of that material or work at the time of bid, e.g., allowance for removal and disposal of asbestos containing materials. Cash allowances must be stated in the Contract Documents and shall be expended as the Owner directs through the Consultant.

As per clause 4.1 of CCDC-2 [1], the contract price includes all cash allowances and in connection with such cash allowances, the contract price, and not the cash allowances, includes the contractor's overhead and profit. Hence, no markup shall be added to any payment made under the cash allowances as it is already covered under the bid price. The value of work performed is included in the progress payments.

FIDIC Conditions of Contract [2] defines "provisional sum" as a sum specified in the contract for the execution of any part of the works or for the supply of plant, material, or services. Sub-clause 13.5 specifies that the Contractor shall be paid for the actual amount used for the work or material and a sum for overhead charges and profit. The provisional sums should only be used for the works as specified in the contract and in accordance with the Engineer's instructions.

12.5.2 Contingency Allowance

A contingency allowance is used for any unforeseen work or costs that may, or may not, ultimately be required. Expenditures from a contingency allowance are administered in the same way as extra cost change orders under the contract. Under Sub-clause 4.2.2, CCDC-2 [1] provides that the contingency allowance includes the Contractor's overhead and profit on expenditures from the contingency allowance. Hence, markup shall be added to any payment made under contingency allowance.

Cost contingencies provide reserves against risk of cost increases during the development of the project. The initial magnitude of the contingencies on any project depends upon many factors, and based on that, contingencies can be established as a specific amount or as a percentage of a budget estimate. The contingencies should be used only on the written instructions of the Consultant to supplement the budget of those operations which have been identified necessary.

12.6 Changes in the Work

The change or variation in work can be defined as changes made in the design, quality, or quantity of the original scope of work set out in the contract. Ontario Provincial Standards [4] defines changes as deletion, extension, increase, decrease, or alteration of lines, grades, dimensions, quantities, methods, drawings, substantial changes in geotechnical, subsurface, surface, or other conditions; changes in the character of work to be done, or materials of the work or part thereof, within the intended scope of the work.

In any construction project, changes to the base scope of work are almost always required as the construction proceeds on site. Most changes are not the result of some fault on the part of a party to the project; they are due to the complexities of construction, unforeseen conditions, changes required by the Owner or the Consultant may need to issue additional information as the work proceeds.

Most of the forms of contract include a clause enabling the Owner to instruct changes in the base scope of the work. As per CCDC-2 [1] clause-6, the Owner, through the Consultant can make changes in the work for any additions, deletions, or other revisions including changes to the contract time by issuing a change order or change directive. It further provides that the Contractor shall not perform a change in the work without having a change order or change directive.

Change order is defined in CCDC-2 [1] as a written amendment to the contract prepared by the Consultant and signed by the Owner and the Contractor stating their agreement upon a change in the work including any adjustment if required to the contract price or contract time. Whereas a change directive is a written instruction prepared by the Consultant and signed by the Owner directing the Contractor to proceed with a change in the work within the base scope of the contract prior to the agreement on any adjustments to contract price or contract time between the Owner and the Contractor.

Similarly clause-13 of FIDIC Conditions of Contract [2] covers the provisions for making changes in the work. However, FIDIC Form allows the Engineer but not the Employer to issue directions to change the works. Each change may include:

(a) Changes in the quantities of any work in the contract
(b) Changes to the quality and other characteristic of any such work
(c) Changes to the levels, positions, and dimensions of any part of the works
(d) Omission of any work under the contract
(e) Any additional work, plant, material, or services necessary for the completion of the works
(f) Changes to the sequence or timing of construction of any part

It further provides that the Contractor shall not perform a change in the work unless and until the Engineer instructs or approves a change.

Per the change provisions of the contract, the Owner and Consultant can amend the works as and when required necessary. In the absence of such a clause, it would be difficult to make some amendments required in the work, and any attempt by the

Owner to initiate changes would technically require a new agreement with the Contractor. The main purpose of the change provisions is to allow such changes to be carried out so as to complete the project smoothly. However, in general, the Owner should not make large scale or significant changes to the nature of the work and beyond the scope of the original contract altogether as it will require a separate contract.

12.6.1 Dealing with Changes

Dealing with changes in a timely and efficiently manner will benefit both the Owner and the Contractor. The early settlement of changes minimizes cost to owners and saves contractors from delaying the project.

Under Sub-clause 6.2, CCDC-2 [1] suggests that when a change in the work is required, the Consultant shall provide the Contractor with a written description of the required work. The Contractor then promptly shall submit the estimate and method for the proposed work with any change in contract price and contract time if required. When Owner and Contractor agree on the submitted quotation, a change order is issued to the Contractor and the contract price is updated accordingly. If the price quoted by the Contractor is not reasonable, then the parties need to negotiate the price with an intention to reach an acceptable price. The value of work performed on the basis of change order is claimed through a progress payment.

The Contractor's request for any change to the scope of work must be consistent with the contract, or it will likely be denied. The Contractor should make changes only when he gets written instructions from the Consultant. If the Contractor acts without an instruction, he will have no grounds to claim for the cost of such change. In order to deal with changes properly, all parties need to be familiar with the contract documents, particularly the conditions covering changes and extras, completion and time extensions, liabilities, and warranties.

Parties to the contract should avoid such actions or inactions which may give rise to a claim, such as the Owner shall avoid delayed approval of Contractor submissions, late design changes, delay in supply of items on behalf of the Owner, failure to coordinate work of third parties, errors or inadequacy of the Contract Documents, and so on. Similarly, the Contractor must assure that the contractual requirements for submittals, reports, and other supporting documentation are met in a timely fashion as the work progresses.

12.6.2 Valuation of Changes/Variations

When a change order for variation is agreed to be issued, the changed or extra work needs to be valued. The rules for valuation of variations suggested in most of the standard forms of contract can be summarized as under:

(a) Where the varied work is of a similar character and executed under similar conditions, the rates and prices set out in the contract should be used to value variations.
(b) Where the varied work is same in character but is executed under different conditions, the rates and prices in the contract shall be used as the basis for valuation, known as pro-rata basis.
(c) If the varied work is different and does not meet above two requirements, then a fair method of valuation is to be used and agreed by all parties, accounting for direct cost, impact cost, and markup (indirect cost or overhead plus profit) as defined in following pages.

For estimating fair value of extra or additional works, the following information could be helpful:

1. For rent of equipment, OPSS127 [7] – schedule of rental rates for construction equipment can be used.
2. For labor rates, statistics Canada provides monthly wages rates for construction union and nonunion workers.
3. Unit prices can also be considered from historical data in previous bids.
4. "Hanscomb Yardsticks for Costing [8]" is also very useful which provides item rates based on every province in Canada.

In the event of disagreement, the Engineer as specified under clause 12.3 of FIDIC [2] shall fix such rates or prices as fairly as possible and in consultation with both parties and notify the Contractor accordingly with a copy to the Employer. Until the rates or prices are agreed or fixed, the Engineer may determine provisional rates or prices so that the Contractor should get his interim payment based on such work done.

12.6.2.1 Time and Material

If the varied work cannot be properly measured or valued, then the Contractor is to be paid for the work done on time and material (T&M) bases. It may be noted that a decision for this method should be made by the Consultant, who should issue such instructions; it is not for the Contractor to insist on valuation according to T&M basis. Payment per this method is calculated based on actual cost of labor hours, material, and equipment used as explained in Chap. 5.

The Contractor should furnish to the Consultant on daily basis the daily work record showing the hours spent on the work, the worker names, plant and equipment employed on the work, and details of material used.

12.7 Construction Cost Control

Different cost terms are used while estimating and controlling project costs, some of those are:

- *Variable costs*: costs that vary or change with the amount of production or the amount of work done. Examples are materials, supplies, and wages.
- *Fixed costs*: costs that do not change with the change in production. The examples are set up cost, rental equipment, or machinery costs.
- *Direct costs*: operating costs except markup (overhead plus profit). Examples are labor, material, plant and equipment, small tools, subcontractor's cost, and testing.
- *Indirect costs*: costs related to head office expenditures. Examples are management and engineering, supervision and inspection [9], insurance and bonds, building rent/utilities, phones, computers, interest and taxes, and other office petty expenses.
- *Impact costs*: costs that are the indirect results or consequences of a change. Examples are disruptions, inefficiencies, delays, accelerations, premium time, remobilization, idle labor, and equipment.

For cost control on a project, the construction plan and the final cost estimate provides a baseline for the assessment of financial performance during the project. For control and monitoring purposes, the original detailed cost estimate is typically converted to a project budget, and the project budget is used subsequently as a guide for management.

For effective cost control, it is important that all three resource parameters – time, cost, and performance – must be analyzed as a group. To be within budget serves no useful purpose if performance is poor. An effective control system monitors schedule and performance as well as cost, measuring expenditures against budgets and identifying variance, assuring that the expenditures are proper and taking corrective action when required. Cost control is a process that should be continued throughout the construction period to ensure that the cost of the project is kept within the agreed cost limits.

An important part of cost control is to calculate the variance, its magnitude and to decide if the variance requires corrective action. There are several cost reporting and controlling systems such as project cost-value reconciliation, overall profit or loss, actual versus forecast reconciliation, earned value analysis, etc.; however, here we will discuss "earned value" reporting system which is the most commonly used method of performance evaluation and project cost control. This reporting system combines performance, time, and budget. The earned value technique measures actual performance of the progress on a project and enables monitoring of both cost and schedule variance.

When using earned value technique (EVT), the following definitions need to be understood:

- *Earned value (EV)*: EV is the budgeted amount for the actual work done within the specified time, i.e., the amount of progress payments (usually monthly) submitted by the contractors.

- *Planned value (PV)*: PV is the planned budgeted cost, required for the work to be done within the specified time. This is calculated on a monthly basis from the baseline schedule of the project. The total amount of activities to be done per month is calculated and recorded to compare it with the monthly progress payments submitted by the Contractor, i.e., the EV.
- *Actual cost (AC)*: AC is the total cost actually paid for the work done within the specified time. This is in fact the actual cost of the activities performed by the Contractor.
- *Cost variance (CV)*: CV is equal to earned value minus actual cost:

  CV = EV − AC and CV% = CV / EV × 100
- *Schedule variance (SV)*: SV is equal to earned value minus planned value:

  SV = EV − PV and SV% = SV / EV × 100
- *Cost performance index (CPI)*: CPI is a measure of cost efficiency. It measures the value of work performed against the actual cost.

  CPI = EV / AC. If above 1, it is considered under budget; if below 1, it is over budget.
- *Schedule performance index (SPI)*: SPI is a measure of schedule efficiency. It measures the value of work performed against the actual cost.

  SPI = EV/PV. If above 1, it is considered ahead of schedule; if below 1, it is behind schedule.

Now to understand EVT for a project, two activities are shown in Table 12.1.

As per activity of Raft Foundation, earned value EV is less than planned value PV, which indicates that the project is behind schedule, whereas the activity of superstructure shows EV is higher than PV, which indicates that the project is ahead of the schedule.

It further shows that AC for Raft Foundation is higher than EV, which means that expenditure is more than the earned value (EV) which indicates that the project is over budget. On the other hand, AC for superstructure is less than the EV which shows that the project is within budget. Similarly the table also shows values for CPI and SPI. For Raft Foundation, CPI is 0.93, which is below 1; that means the activity is over budget. For superstructure, CPI is 1.15, which is above 1 meaning the activity is under budget. Similarly for Raft Foundation, SPI is 0.92, which means it is behind schedule, whereas for superstructure the SPI value is 1.10, which means above 1: hence, it is ahead of schedule.

The negative variance for Raft Foundation clearly indicates that there is a loss; hence, corrective measures are immediately required. The positive variance is an encouraging sign that indicates that amount of work done at the specified time is more than the work planned on that time; hence, the project is ahead of the schedule.

Table 12.1 Example of cost and schedule variance

Performance report sample

	Budget	Earned value	Actual cost	Cost variance		Schedule variance		CPI	SPI
WBS element	PV ($)	EV ($)	AC ($)	EV−AC ($)	CV/EV (%)	EV−PV ($)	SV/PV (%)	EV/AC	EV/PV
Raft foundation	63,000.00	58,000.00	62,500.00	−4,500.00	−7.75	−5,000.00	−7.90	0.93	0.92
Super structure	200,000.00	220,000.00	190,000.00	30,000.00	13.00	20,000.00	10.00	1.15	1.10

WBS = Work breakdown structure

12.8 Forecasting

Forecasting is also one of the tools to monitor and control costs. Earned value technique also provides formulae to forecast the future performance of the project.

EAC: Estimate at Completion The expected total cost of completing project work.

It is the forecasted cost of the project as the project progresses. There are different methods to determine EAC; however, in its most commonly used form, it is the budget at completion (BAC) divided by the cost performance index (CPI), i.e.,

EAC = BAC / CPI (BAC is the original total budget allocated to the project).

This method is used when we assume that future work will continue at the present cost performance index, meaning the change encountered by the project is typical.

Other methods for calculating EAC include:

EAC = AC + ETC. This is called a "bottom-up" formula.
EAC = AC + (BAC − EV). This method is used when the project has encountered a one-time (atypical) change.

ETC: Estimate to Complete The estimated cost of completing the remaining work.

ETC is a forecast of how much more money will need to be spent to complete the project.

There are primarily two methods of calculating ETC:

ETC = (BAC − EV)/CPI

Similar to EAC method, this method is used when we consider that future work will continue at the present cost performance index, meaning the change encountered by the project is typical.

ETC = EAC – AC. This method is used when we have calculated EAC.

Simple Example 12.1
Now as a simple example to understand difference between EAC and ETC, let us consider the activity of Raft Foundation within Table 12.1 above; we will have:

EAC = BAC / CPI = $63,000.00 / 0.93 = $67,741.93 (overall estimate on completion)
ETC = (BAC − EV) / CPI = ($63,000.00–$58,000.00) / 0.93 = $5376.34 (estimate for remaining amount to complete the project).

Once we know EAC and ETC, we can calculate variance at completion (VAC).

VAC is the difference between the original project cost (baseline) and what it is currently expected to cost.

VAC = BAC − EAC = $63,000.00 − $67,741.93 = −$4741.93

Negative variance indicates over budget, meaning the project lost money.

12.9 Value Engineering

Value engineering is one of many cost-saving techniques. It is an organized approach to identify and eliminate unnecessary costs. It helps with elimination or modification of anything that adds cost to an item without contributing to its required functions. During this process, all expenditures relating to design, construction, maintenance, operation, replacement, etc. are considered.

Under value engineering, the functions of a product or service or project activity are analyzed in order to reduce the overall cost, to determine "best value" or the best relationship between worth and cost, without sacrificing quality or performance requirements. It usually involves steps of gathering required information, searching for creative ideas, evaluating suitable alternatives, and proposing cost-effective alternatives. The major objective of value engineering as applied to construction projects is to reduce initial and life cycle costs.

At the planning and design stage, designers design one or more systems to satisfy the requirements and then select a system for value analysis. Next, the designers should question whether the system chosen provides the best value at the lowest cost. Value engineering is a useful procedure for answering this question and selecting a better alternative if the answer indicates this is desirable. At the planning and design level, value engineering is mostly done in four stages:

1. *Information gathering stage*: This stage involves in identifying the main elements of a product, service, or project and analyzing the functions of those elements. Function analysis, an important technique in value engineering, is usually done in this stage. At this stage, key criteria and objectives for the project are defined. The objective of information gathering is to determine what functions or performance characteristics are important. Questions need to be asked like: What is the purpose of the project? What should it provide? What should it not include? And so on.
2. *Alternative generation stage*: In this stage, alternative solutions are developed for delivering necessary functions. Various alternative ways of meeting requirements are created and reviewed for additional ideas that would perform the desired function.
3. *Analysis stage*: In this stage, all the alternatives are assessed by evaluating how well they meet the required functions and how successfully expenditure is minimized. Ideas found to be impractical are discarded. Ideas with greatest potential for cost savings and value improvement are considered.
4. *Decision stage*: In the final stage, each recommendation is presented with a brief narrative to compare their advantages and disadvantages. Wherever required, sketches and design calculations are included in this stage along with cost comparison and life cycle cost calculations. Finally, the best alternative is to be chosen and proposed to the Owner.

The value engineering efforts should be initiated at the early stages of project design because early review results in greater savings and allows a change of direction, if appropriate, without affecting project delivery schedules. Note that substantive change in later stages incurs more costs to parties involved due to duplicated work or rework.

During the construction phase, contractors are often required to suggest changes in the plans or specifications and to share in the resulting savings. These changes may involve substitution of materials, modification of design, reduction in quantities, or procedures other than those set forth and required by the Contract Documents. Value engineering is designed to take advantage of the Contractor's special knowledge and to cut the cost of a project to the lowest practicable level without compromising its function or sacrificing quality or reliability. In short, the Contractor is encouraged to develop and submit to the Owner cost-reducing proposals leading to changes in the plans or specifications. If the proposals are accepted by the Owner, a change order/variation is processed, and the savings are usually shared equally by the Owner and the Contractor. Contractors are willing to examine such alternatives when offered incentives for sharing the savings by the Owner.

Another form of cost savings from value engineering is the expertise of Contractor's site staff to take advantage of such techniques which affect the cost of various activities/items/functions performed without sacrificing quality. For example, the depth of a strip footing is 300 mm, whereas the depth of most of the floor beams is 450 mm; hence, it is advisable to order formwork of 450 mm, so as to reuse the same formwork in foundations as well as throughout the upper floors. Similarly, using steel formwork can provide cost saving with a multistory building instead of wooden formwork. The wooden formwork may lose its strength after reuse four or five times, whereas steel formwork will remain mainly unaffected.

Another example could be use of precast staircases instead of cast in situ, in a high-rise building. With a precast staircase, there will be more saving by cutting the cost of reusing formwork, excessive manpower, steel fixing, and concreting process throughout in the upper floors of a high-rise building. In such a way during construction, a sensible Construction Manager can reduce the overall cost of the project and increase his profit by utilizing his expertise by proposing various alternatives in performing construction activities.

12.10 Life Cycle Costing (LCC)

Life cycle costing or whole life costing is also one of the cost control and cost awareness techniques which takes into account the total cost, both present and future. It can be defined as an economic evaluation technique that involves the assessment of the total cost of an asset over its operating life, including initial capital costs, maintenance costs, operating costs, and the cost or benefit of the disposal of the asset at the end of its life.

The LCC analysis assists funding decisions by highlighting the total cost of an asset over the course of its life expectancy, rather than focusing on initial construction costs only. It can also be used to compare investment alternatives with different initial and future costs. For example, for an infrastructure asset it can be used to compare construction and material costs of asphalt pavement versus concrete pavement including assessment of future cost of operating, maintaining, rehabilitating, and replacing an asset.

The major benefits of LCC analysis can be summarized as:

1. Improves awareness of total costs
2. Gives a picture of true total costs
3. Identifies choice of alternative courses of action
4. Identifies those areas where costs may be reduced
5. Allows to arrange for future funds

A clear understanding of life cycle phases for a project permits managers and executives to better control total corporate resources in the achievements of desired goals. The life cycle phases of a project may include:

(a) *Initiating or conceptual phase*: This includes the preliminary evaluation of an idea with assessment of the technical and financial viability of the project. The most important component in this phase is a preliminary analysis of risk and resulting impact on the time, cost, and performance requirements, together with a potential impact on company resources.
(b) *Planning or definition phase*: This phase includes identification of work to be done with initial preparation of all documents like schedule of works, quality control procedures, etc. This phase indicates planning methods that are applicable and provides the basis for controlling an operation in order to reach an established goal. It includes appointment of the design team and other consultants, establishment of budget including cash flow, identification of the appropriate contract strategy, and determining bidding procedures and select Contractor.
(c) *Construction or production phase*: It includes executing, controlling, and closing processes. It involves monitoring of progress and expenditure, direction and inspection of works, certification of payments, checking safety and quality control systems, approval of variations, holding of progress meetings, and problem resolution.

    Finally, during closing process, this phase includes issuing of completion certificates, ensuring correction of deficiencies, ensuring final accounts are settled, arranging As-Built drawings, commissioning, and handing over.
(d) *Operational phase*: This includes the operational life of the project and provides evaluation of the technical, social, and economic sufficiency of the project to meet actual operating conditions.
(e) *Divestment phase*: It includes disposal, termination, or replacement of the project.

Once the life cycle phases of the project are identified and understood, the cost breakdown structure (CBS) is prepared. The aim is that the whole life cost model should include every cost likely to be incurred in respect of the project from inception to disposal. The CBS represents the way LCCs are broken down and presented.

Each cost element within the CBS must be well defined so that all involved have a clear understanding of what is to be included in that element. The CBS should be designed to allow different levels of data within various cost categories according to their functionality. After the production of a CBS, costs for each category are calculated. Cost estimates are usually made up of the base estimate and risk allowance.

Reference and Further Reading

1. CCDC-2. (2008). Canadian Construction Document Committee. Stipulated price contract.
2. The FIDIC. (1999). Conditions of contract for construction – (Red book). Author FIDIC.
3. Ontario Construction Lien Act R.S.O, 1990, Chapter C30- Last Amended 2010,c.16,sched 2,s.2- Consolidation Period: From July 1, 2011 to e-Laws currency Date Nov 26, 2013.
4. OPS. (2006). General conditions of contract- Ontario provincial standards for roads & public works- OPSS. MUNI 100.
5. CCDC. (2001). Canadian Construction Document Committee – Statutory Declaration forms 9A.
6. Takeover Procedures- Ontario Association of Architects and Ontario General Contractors Association (OAA/OGCA) – Document # 100.
7. Ontario Provincial Standard Specification- OPSS 127. (2015). Schedule of rental rates for construction equipment, including model and specification reference.
8. Hanscomb. (2014). Yardsticks For costing- Cost data for the Canadian construction industry. Published By RS Means.
9. Levin, P. (1998). *Construction contract claims, changes & dispute resolution*. Reston: ASCE (American Society of Civil Engineers) Press USA.

Chapter 13
Construction Quality

13.1 Introduction

Quality of work, along with time and cost, is some of the most important factors to be considered in construction of any project. Defects or failures can result in huge additional expenditure and delays. As per a recent survey on Quality of Construction by FIDIC within Member Associations in 2001, it was found that quality was frequently adversely affected due to a reduction in the initial costs of construction and supervision [1]. In fact, the contractors and consultants both usually submit low prices to win the tenders but at the risk of being unable to produce and achieve quality work meeting the standard specifications. Lack of quality in construction leads to poor workmanship, and in delays, cost overruns, and disputes in construction contracts.

Generally, the Contractor is bound to execute and complete the work with due care and diligence in accordance with the provisions of the contract and required standards. The Contractor is responsible for means, methods, techniques, sequences, and procedures of construction. Whereas, the design professional is obligated to protect public health and safety in the context of the final completed project. The required standards of quality of a completed project are made during the design and planning stages rather than during construction.

Although quality management is important at every stage of a project life cycle, quality management during the construction stage contributes significantly to the final quality outcome of a construction project. This chapter mainly focuses on quality control and quality assurance required during the construction phase. Quality management in construction typically involves maintaining compliance with required standards of material and workmanship in order to insure the performance of the facility according to the design. These required standards are outlined within the specifications and drawings of the contract.

A. Surahyo, *Understanding Construction Contracts*,
DOI 10.1007/978-3-319-66685-3_13

13.2 Contract Requirements

As per FIDIC [2] Sub-clause 7.1, the Contractor shall carryout the manufacture of plant, materials, and production and execution of work in conformity with the contract and in accordance with recognized good practice. For the purpose of insuring compliance, the Engineer or his staff arranges tests of the materials used from time to time, as the basis for accepting or rejecting work completed and batches of materials. As per FIDIC [2] Sub-clause 7.5, the Engineer has the authority to issue instructions to the Contractor for removal from the site such materials or plants, which in his opinion are found defective or not in accordance with the contract and to replace the same at his own cost.

Similarly CCDC-2 [3] also specifies in Sub-clause 2.2.12 and 2.4.1that the Consultant has the authority to reject work which does not conform to the requirements of the Construction Documents, and the Contractor shall rectify the defective work promptly. CCDC-2 [3], Sub-clause 3.1.1 also provides that the Contractor shall have total control of the work and should direct and supervise the work in conformity with the Contract Documents.

With conformance as the measure of quality during the construction process, the quality requirements specifications within the Contract Documents become extremely important. Quality requirements should be clear and verifiable so that the parties in the project can understand the requirements for conformance.

13.3 Quality Management

Various definitions of the term "quality" are used by different stakeholders. Some consider it to be the characteristics of a product or service that bear on its ability to satisfy a need; others take it as the degree to which a product exceeds a customer's requirements and expectations, whereas some consider it conformance to requirements. However, Schexnayder and Mayo [4] have added that construction quality is also about finishing the project safely, on time, within budget, and without claims and litigation.

Quality management includes all those activities, which are performed to ensure that the project will satisfy the needs for which it was undertaken. Quality management in construction provides many benefits such as:

(a) Helps avoid defects and rework
(b) Reduces waste of project resources
(c) Lowers costs
(d) Affords greater certainty of project success
(e) Higher productivity
(f) Stronger customer loyalty and satisfaction

Sometimes, quality is sacrificed to save time and reduce costs, whereas quality itself saves time and money. It is therefore common saying that "nothing saves time and money more than doing work right the first time."

13.4 Quality Management Processes

A Quality Management Process is a set of procedures which are being followed to make sure the product achieved is fit for purpose and meets the quality standards spelled out in the Contract Documents. Under this process a variety of review techniques are undertaken, and required corrective actions are implemented to address any deficiencies to achieve the proposed quality levels within the project.

Quality management on a project is based on processes such as quality planning, quality control, quality assurance, and QA/QC implementation.

13.4.1 Quality Planning

The overall objective of the quality plan is to ensure the project is constructed safely, efficiently, and economically in accordance with the plans and specifications under the contract. In quality planning, it should be identified which quality standards are to be used in the project and how they will be implemented. Standards are defined by the American Society for Quality (ASQ) in its *Quality Engineering Handbook* [5] as "documents used to define acceptable conditions or behaviors and to provide a baseline for assuring that conditions or behaviors meet the acceptance criteria." Standards identify a bench mark for material specifications or test methods and installation practices. They state the test methods by which something is to be made rather than the results to be achieved.

To achieve quality objectives, these required standards should be defined at the design stage by the consultants in the specifications and drawings of the project. Without early identification of the appropriate codes and standards, reworking plans and specifications can result in considerable cost and delay. The plan should define clearly the acceptance and rejection criteria for the material to be incorporated and work produced. Inconsistencies between the drawings and specifications should be avoided. The specific procedures, along with a method of variation and addition or alteration required during the course of construction should also be defined.

The Contractor is responsible for establishing a quality control plan after award of contract, demonstrating how the Contractor's responsibilities for quality control will be fulfilled ensuring that work complies with the requirements of the Contract Documents. The minimum Contractor quality control activities are defined in the construction contract. The Contractor's quality control plan should define:

1. Resources to be used.
2. Personnel responsibilities, authorities, and their training for the work.
3. Inspection procedures to ensure effective quality control and assurance of quality for acceptance of materials and workmanship. This includes but is not limited

to fabrication, sampling and testing, production, storage, delivery, construction, and placement.
4. Methods to be used to identify nonconformities and implement corrective and preventive actions.
5. Methods to be used for document control and records management, maintaining a record of inspections, including but not limited to, date of inspection, results of inspection, and any subsequent corrective actions taken.

13.4.2 Quality Control (QC)

Quality control in construction can be defined as, "The product resultant from successful executions, using operational techniques and activities by the Contractor satisfying that the required standards or implied needs of quality construction are met." Since QC is a production tool, it is therefore part of the contractors program to assure compliance with the quality requirements mentioned in the Contract Documents. The primary function of QC is the successful execution of a realistic plan to ensure that the required standards of quality construction will be achieved.

For effective QC, the Contractor defines procedures to manage and control his own forces, his subcontractors, and his suppliers' activities so that the completed project complies with all contract requirements. A well-defined QC program should be established for the project. Designers should work closely with contractors to meet project quality objectives. While carrying out quality control, the Contractor has to ensure that standards are communicated to the subcontractors and that the work undertaken complies with those standards.

An effective program is essential and beneficial to the Contractor. The savings from avoiding replacements, repairs, reworking, claims, and litigation will exceed many times over the cost of an effective QC program. In order to deliver a successful project, it is essential for the construction company to take the following steps before any work is started:

1. Have the specific objectives clearly defined.
2. Ensure that the teams have necessary resources and organizational support to carry out their designated functions.
3. Select the entire team early to use their expertise from the inception of the project.
4. Have a total understanding of the quality control that will be enforced, as well as the complete testing program to be used throughout the construction period.
5. Make it clear that management is committed to quality.
6. Raise the quality awareness and personnel concern of all employees.
7. Make sure that everyone works effectively for the company.

13.4.3 Quality Assurance (QA)

The primary function of quality assurance is to obtain completed construction that meets all contract requirements. Assurance is defined as a degree of certainty ensuring that the final product will satisfy given requirements for quality. Quality assurance is a method for the regular monitoring and evaluation of the various aspects of a project to ensure that standards of quality are being met. Quality assurance personnel monitor the Contractor's work to ensure that the work complies with the Contract Documents. Since quality assurance is a management tool, it is mostly part of the Owner's responsibility.

An effective QA program is beneficial to the all parties. With an effective QA program, the Owner will have more cost control, fewer claims, improved constructability, and faster project delivery with minimum repairs. The lack of an effective QA program will have significant long-term costs for the Owner. QA procedures are established by a Consultant and Owner and are put in place prior to execution of the work. The Consultant and Sub-consultants carry out quality assurance by means of establishing quality standards in the drawings and specifications, reviewing the job as it progresses, and requiring the necessary testing to be done by third-party companies.

The elements of an acceptable QA program include quality control by the Contractor, testing and inspection, use of authorized laboratories, and the use of qualified personnel. The Contractor carries out quality control by working with the Consultant to meet the specified quality standards, making sure that the standards are communicated to the subcontractors, and that the work undertaken complies with those standards ensuring that deficiencies are brought to light as early as possible and addressed immediately by the responsible trade.

The supervision exercised by the Consultant representative on site must ensure that the materials incorporated, machinery used, and workmanship provided by the Contractor are of the required quality. In order to implement this process of quality in construction, the following points should be considered before starting the project [6]:

1. Appoint the project management/quality assurance team early with clearly defined responsibilities.
2. Develop a QA system for the project. Indicate which section of the QA team shall develop the QA plan and QA programs at various stages of planning, design, and selection of materials, construction supervision, and material testing.
3. Establish who does what within the organization.
4. Select the people who are qualified for the particular tasks and train the personnel in their responsibilities.
5. Conduct preconstruction meetings to ensure that everyone understands his contract responsibilities and all necessary activities. A preconstruction meeting is important for both the QC and QA programs for smooth running of the project. It establishes cooperative and non-adversarial relations between both parties.

During the meetings, both programs should be reviewed and the lines of authority and communication established and recognized.
6. Monitor and review the progress of the teams regularly. This will facilitate alterations to the strategy at an early stage, if needed.

13.5 QA, QC Implementation

Quality assurance and quality control are extremely important aspects of any project, without which, successful completion of the project will not be satisfactory. The requirements of a comprehensive QA/QC programs should be defined in the quality plan at the project planning stage. The quality plans should cover the type of test, test standard, frequency, control requirements, and assigned responsibility for inspections and tests. The implementation of QA/QC systems should be organized on site in such a way that adequate and continuous control is exercised over all the activities affecting quality. Proper implementation of quality assurance and quality control not only results in a sound project but also leads to more economical and time-saving outcome.

For successful implementation of QA/QC programs, the activities reviewed and monitored on site during construction include:

1. Contract requirements for quality control and acceptance of material and workmanship are enforced.
2. Standard inspections and testing are performed.
3. Proper inspection techniques such as checklists are used.
4. Adequate document control is in place.
5. Requirements of the specifications and submittals are properly implemented.

13.5.1 Quality in Materials

Materials incorporated into the construction directly affect the quality of finished work. In general, the Contractor's quality control is designed to monitor, assess, and adjust the production or placement processes of specific materials to ensure that the final product will meet the specified quality level. The quality requirements for materials are contained in the specifications and other Contract Documents including with Contractor's submittals. These documents include the acceptance criteria for determining if the prescribed quality for materials has been met.

To avoid any complications, the Contractor should make sure that the requirements for materials supplied to site are clearly defined and communicated to the manufacturers and suppliers. Once the Contractor finalizes his selection for suppliers and materials to be used, he should submit the material data for approval of the Consultant. The review of submittal data is one of the first steps on site in the QA/

QC process. The information received from the Contractor, subcontractors, and suppliers for items to be incorporated into the project must be verified to meet the standards set forth in the Contract Documents. The request submitted by the Contractor for incorporating any material should indicate the following:

1. Name of material
2. Location, where the material will be used, e.g., foundation, etc.
3. Manufacturer's name
4. Name of the local supplier
5. Contract specification reference
6. Manufacturing standard
7. Submittals like: samples, test results, catalogues, mill certificates, shop drawings, etc.

The above data, for any material to be used, should be reviewed and verified by the Consultant carefully. Checking submittal information becomes more important when shop drawings are required. Shop drawings are usually required when the contract drawings do not provide enough detailed information, like fabricated works, elevators, bar-bending schedule, rolling shutters, etc.

Once the submittal data found meets the contract requirements, the Consultant can give approval for the required material, and the sample and record can be stored for verification and future reference for supply of materials to the site. If the data is deemed as "rejected" or "incomplete," the Contractor should resubmit another sample or furnish more information for re-approval. Any mistake not discovered in the submittal review process will lead to potential problems involving extra cost and additional time for rectification.

The Kansas City Hyatt Regency walkway collapsed in 1981 is an example of how a poor shop drawing review can lead to disastrous consequences. In this case, a change in the details of structural connections, left unchecked during the submittal process, doubled the load on the fourth floor walkway connections. This extra load on these connections lead to the collapse of the fourth floor suspended walkway onto the second floor walkway and then onto the ground floor below. This resulted in 114 people being killed and over 200 people injured. In addition, millions of dollars in costs resulted from the collapse, and thousands of lives were adversely affected.

13.5.2 Quality in Workmanship

Poor construction methods and workmanship can be responsible for the failure of any project. Poor construction methods and workmanship are mainly resultants from negligence and inadequate quality control at the construction site. In general, the Contractor's obligation is to carry out and complete the works in a proper and workmanlike manner as shown on the Contract Documents. In addition to the Contractor's own quality control measures, site inspectors working as Consultant's representatives inspect the works to verify compliance with the requirements of the Contract Documents.

The quality requirements are usually specified in the plans and specifications. Planning for quality in workmanship should ensure that the construction activities proceed under controlled conditions in the specified manner and sequence. One objective of a good QA and QC program is to prevent errors from occurring or to find errors quickly after they have occurred. In such cases, it is especially important to have an inspection and testing program for work in progress.

Inspection process Inspection is one of the most important aspects of construction work. It involves monitoring the Contractor's construction processes to ensure that the construction quality and workmanship are in compliance with the plans and specifications.

Inspections for received materials on site should ensure that they conform to the approved submittal data or sample stored at the site. For this purpose, a checklist for each item may be prepared to verify incoming material meets contract requirements and approved submittal data. After inspection, if all information is in agreement, then the materials can be allowed, otherwise they should be rejected and removed from site to prevent unqualified supplies from being inadvertently used. Inspection reports should identify observations conducted, results of inspections, location and nature of defects found, causes for rejection, and remedial or corrective action taken or proposed. Additional QA inspections may include inspection of third-party lab testing facilities, fabrication facilities, and suppliers.

For inspection, the verification of setting-out, orientation, elevation of work, and any embedded material required is extremely important. Work that is not placed correctly will result in an extra cost for rework. It is most important that qualified, experienced, and trained personnel should carry out the inspection works. For any work inspection, it is preferable to have a checklist. The checklist will vary, based on the nature of work to be inspected. Before beginning a work activity, the site must be inspected to determine readiness of the work area.

Testing process Testing is the most important part of the QA and QC process. All tests during construction should be performed as per the Contract Documents. As required by the contract specifications, the Contractor under the QC program shall establish a test program for those tests which are to be conducted by the Contractor to ensure that all required testing is properly identified, planned, documented, and performed.

QA inspection and testing will be used to verify the adequacy and effectiveness of the Contractor QC program. Under the QA program, the Owner usually employs an independent inspection company to carry out visual inspection and destructive/nondestructive testing as deemed necessary for some of the following sections of work:

1. Compaction of backfill and granular base courses
2. Concrete slump and air test, cylinder compression test, and aggregate analysis
3. Report on field and shop welding of steel and stainless steel pipe and equipment

4. Report on field welding and high-strength bolt connections for structural steel
5. Load tests of structural items
6. Subgrade examination for load-bearing capability if required
7. Pile testing operations
8. Asphalt pavement tests

All testing should be performed by qualified personnel and through an approved laboratory or at site, if proper testing equipment is available. The record of test results should be maintained properly.

The test results should be evaluated by a qualified person to ensure the testing meets the acceptance requirements mentioned in the Contract Documents. Any non-conforming test results should be brought immediately to the attention of the Consultant and the Contractor. It may be noted that non-compliance by a single test does not necessarily mean that the item is inferior to that specified; the Consultant should determine its adequacy by evaluating the test result further or investigating by taking more tests.

Some of the common tests that are being performed on the construction sites to ensure the quality of work being performed are:

(a) *Soil compaction test*: The soil is usually compacted prior to laying any structure so that the weight of the structure (building, road, or embankment) should be distributed over a large area to avoid overloading of the soil beneath and also to prevent unequal settlement. Most of the codes require that the soil should be compacted minimum to 95% of the maximum dry density. Soil samples are collected from the site by lab technicians, and the maximum density is calculated in the laboratory. Nuclear density gauges are used to measure in-place density of soils.

(b) *Concrete slump test*: A slump test is performed to ensure that concrete being placed is neither too wet nor too dry. The test is performed in a cone of 300 mm (12 in) height, 200 mm (8 in) diameter at the base and 100 mm (4 in) at the top. The cone is filled by concrete in three layers, rodding each layer 25 times with a standard 16-mm (5/8 in) slump rod. The slump cone is then lifted up slowly and placed up turned near the compacted concrete. The depth between top of up-turned cone and concrete is measured, which is known as slump.

Ontario Provincial Standard Specification (OPSS) – 1350 recommend following slumps based on slump components prior to the addition of super-plasticizer [7].

1. Reinforced concrete within vertical formwork such as abutments columns, piers, walls, beams, and footings: 80 ± 20
2. Reinforced or plain concrete in flat sections such as bridge decks and approach slabs: 70 ± 20
3. Slip-formed barrier walls: 15 ± 10
4. Tremie Concrete: 150 ± 20 (Revised OPSS in 2013 allows max: slump 180 mm)

Based on slump test results (without addition of super-plasticizer), degree of concrete workability is often described in four categories as shown in Table 13.1:

Table 13.1 Categories of concrete workability

Degree of workability	Slump		Category
	mm	in	
Very low	0–25	0–1	Very dry mixes
Low	25–50	1–2	Low workability mixes
Medium	50–100	2–4	Medium workability mixes
High	100–175	4–7	High workability concrete

Maximum slump for concrete to which super-plasticizer has been added is recommended by OPSS – 1350 as 180 mm.

(c) *Concrete cube or cylinder test*: Concrete cubes or cylinders are prepared on site and then they are usually taken to the laboratory for testing purposes to ascertain the compressive strength of the concrete actually placed. Primarily, concrete cubes or cylinders are the two types of compressive test specimens used to determine the compressive strength. Standard cubes used on site are usually 150 mm × 150 mm (6 in × 6 in) size, whereas cylindrical test specimens used of various sizes are 100 mm × 200 mm (4 in × 8 in) or 150 mm × 300 mm (6 in × 12 in). The required specimens are casted on site in three layers, rodding each layer 25 times with a standard rod. While making cylinders or cubes for acceptance of concrete, other properties of the fresh concrete are also checked, such as concrete temperature, slump (as mentioned above), density (unit weight), and air content. The casted specimens are stored and cured in the field until the concrete hardens and are then taken to laboratory for testing after 7 days and 28 days. In the laboratory, tests are performed by crushing these cubes in a standard machine. A strength test result is always the average of at least two specimens tested at the same time; therefore, two cubes or cylinders shall be casted for crushing on the 28th day.

Concrete usually gains strength over a long period of time; hence, the compressive strength at the age of 28 days is commonly used as measure of this strength. The 7-day tests are a guide to the rate of hardening.

(d) *Testing of water*: Samples of water used for mixing and curing concrete are collected from site randomly during construction and sent to the laboratory for testing. Water used in concrete for mixing and curing should be of such quality that it should not affect the setting time, strength, durability, reinforcement, and appearance of the hardened concrete. For these requirements, water used should be clean and free from objectionable quantities of organic matter, silt, clay, acids, alkalies, and other salts and sewage.

CSA (Canadian Standards Association) – A23.1-09/A23.2-09 [8] specifies that any potable water is suitable for use in the manufacture of concrete. Whereas, non-potable water may be used for concrete production provided that a satisfactory history of strength and durability of concrete made with the water has been demonstrated.

Mostly the tests required for water to be used for mixing and curing concrete are to assess the pH value and contents of total dissolved solids (TDS), sulfates, alkalies, and chlorides. Mostly the limitations for these substances in water are:

pH value	6.5–8.0
Total dissolved solids (TDS)	Less than 2000 ppm (mg/lit)
Sulfates as SO_3	Less than 1000 ppm (mg/lit)
Alkalies HCO_3 /CO_3	Less than 1000 ppm (mg/lit)
Chlorides as NaCl	Less than 800 ppm (mg/lit)

(e) *Coarse and fine aggregates*: Like water, samples of aggregates are also collected from site randomly during construction and sent for laboratory testing. The size and grading are important properties of aggregates as they affect the workability and strength of the concrete mix. The proper grading of aggregates will demand a lower water-cement ratio that results in an increase in strength. It helps prevent segregation during placing and ensures a good finish.

The aggregate size also affects the strength. For given mix properties, the concrete strength decreases as the maximum size of aggregate is increased. Sieve analysis is conducted by laboratory on the samples collected from site, and the results are matched with the size and aggregate grading requirements specified in the contract to approve or reject. Chemical analyses are also sometimes required, which should meet the following criteria:

- Coarse Aggregate:
    1. Maximum permitted content of acid soluble chloride should be 0.05% by weight of aggregate.
    2. Maximum permitted content of SO_3 should be 0.40% by weight of aggregate.
- Fine Aggregate/Sand:
    1. Maximum permitted content of acid soluble chloride should be 0.10% by weight of aggregate.
    2. Maximum permitted content of SO_3 should be 0.40% by weight of aggregate.

(f) *Fill materials*: Usually there are many classifications of fill materials used for road or building works. However, for convenience, we will divide them as granular fill and sand. These materials are collected from site for testing purposes to know their plasticity index (P.I), liquid limit (L.L), sulfate content, and sieve analysis.

As recommended by various codes, for road base coarse, granular material and sand should be non-plastic, whereas, for subbase and other fillings, the L.L shall not be more than 25%, and P.I shall not exceed 8% for a granular fill material.

As regards to sulfate content as SO_3, it shall not exceed 1%, and TDS (total dissolved solids) shall not exceed 2%. The grading analysis, if carried out for fill materials, should meet the requirements of the grading envelop (or gradation) as mentioned in the specifications.

Generally, fill materials that are highly plastic as indicated by higher value of P.I are highly compressible.

(g) *Leak testing of water mains*: All waterlines inside building and supply lines outside building, i.e., underground, are tested for leaks to make sure that the work is done properly. Testing for leaks involves subjecting all pressurized pipes to hydrostatic pressure testing, which is measured by a water pressure gauge. When the installation within the building is complete, it is slowly filled with water, with the highest draw-off point open to allow air to be expelled from the system.

The pipes, pipe fittings, and connected appliances are then subjected to a test pressure of about 7 bars (about 100 psi/690 KPa) or 1.5 times the maximum working pressure. This pressure is applied and maintained for about 2 h, and the installation is then checked for any loss of water or visual evidence of leakage. If leaks are found, the defects shall be rectified then the system is retested until satisfactory results are obtained.

External underground water mains other than polyethylene pipes are tested in sections between valves or often the completed water main is tested as one. The openings are closed temporarily in the pipeline to be tested. The testing should be carried out before the backfilling of joints; however, other water-main sections can be backfilled by one or two layers to ensure pressure doesn't blow apart the connections. The pipeline is slowly filled with water so that all air is expelled and then tested under pressure. Pipe sections shall be tested at a test pressure of 1035 KPa (10 bars or 150 psi) as per Ontario Provincial Standard Specification (OPSS) – 701 [9].

OPSS 701 [9] further suggests that a 24-h absorption period may be allowed before starting the test. The test section shall be subjected to a continuous test pressure for 2 h. The leakage is then calculated for the section under test and compared with the allowable leakage. The allowable leakage is 0.082 l/mm of pipe diameter per km of pipe for the 2-h test period. If the measured leakage exceeds the allowable leakage, all leaks shall be located and repaired, and the test section shall be retested until a satisfactory result is obtained. In order to avoid the risk of contamination, many standards suggest to use potable water for testing purposes.

Polyethylene pipe Testing of polyethylene pipe is a little different than other pipes. The test procedures consist of two steps: The initial expansion and the test phase. When the test pressure is applied to a water-filled pipe, the pipe expands. During the initial expansion phase, the test section shall be pressurized to the test pressure and sufficient makeup water added each hour for 3 h to return to the test pressure. After the initial expansion phase, the test phase begins.

As per OPSS 701 [9] recommendations, the test phase shall be 2 h after which a measured amount of makeup water is added to return to the test pressure. If the

Table 13.2 Test phase makeup amount for pressure polyethylene pipe [9]

Pipe diameter (mm)	Makeup water (Liter/Km)
30	12.38
40	12.38
50	13.62
75	18.6
100	31.0
150	74.5
200	124.2
250	161.4
275	248.3
300	285.6
350	335.2
400	409.7
450	533.9
500	682.9
550	869.1
600	1105.0
700	1378.2
800	1775.5
900	2234.9
1000	2731.6
1050	3104.0
1200	3973.2
1350	5152.7
1600	7449.7

amount of added makeup water does not exceed the values provided by Table 13.2, leakage is not indicated. If the amount of makeup water exceeds those values, all leaks shall be located and repaired, and the test section shall be retested until a satisfactory result is obtained. Recommended makeup amount of water by OPSS 701 for 50 mm, 100 mm and 200 mm pipes is 13.62 L/Km, 31.0 L/Km and 124.20 L/Km, respectively. For further pipe diameters, OPSS 701- Table 13.2 in following pages can be referred.

The test duration should not exceed 8 h. If the pressure test is not completed, the test section shall be depressurized and allowed to relax for at least 8 h before bringing the test section up to pressure again. As per the Plastic Pipe Institute, air testing is not recommended. Additional safety precautions may be required.

Flushing and Disinfecting Water Mains After installation and pressure testing, new water mains should be disinfected according to OPSS 701 [9] and American Water Works Association (AWWA) – C651 [10].

OPSS 701 [9] recommends that the water main shall be flushed to achieve a minimum velocity of 0.76 m/sec; otherwise the water main shall be swabbed. After flushing is completed, water may be allowed to flow at a controlled rate into the new pipeline. Liquid chlorine solution shall be introduced so that the chlorine is distributed throughout the section being disinfected. The chlorine concentration applied shall be 50 mg/l minimum throughout the section. The system shall be left charged with the chlorine solution for 24 h.

Sampling and testing for chlorine residual will be carried out by the Contract Administrator. The chlorine residual will be tested in the section after 24 h. If tests indicate a chlorine residual of 25 mg/l minimum, the section shall be flushed completely and recharged with water normal to the operation of the system. If the test does not meet the requirements, the chlorination procedure shall be repeated until satisfactory results are obtained.

Twenty-four hours after the system has been recharged, the Contract Administrator will take samples for bacteriological tests. If there is indication of contamination, the disinfection procedure shall be repeated. The disinfection chemicals should be limited to less than 12% active chlorine. The duration of the disinfection should not exceed 24 h. Upon completion, the system should be thoroughly flushed with freshwater and retested to verify the disinfectant chlorine level has been reduced to potable drinking water concentrations in all mains, services, and branch lateral pipes.

(h) *Leak Testing of Drainage/Sewer System*: The drainage line inside a building is tested by filling with water to create internal pressure of about 3.0 m head [11] above the invert of the pipe as recommended by Ontario Building Code. This water is left for about 15 min, and any drop in the level of water in the pipe confirms the leakage. Testing is usually carried out in sections between inspection chambers or other suitable points of access and through any accessible branch drains. Sometimes test branches are frequently extended up to the finish floor level with a suitable termination and used as an additional point of access.

For external sewer lines, leakage tests (infiltration, exfiltration with water, exfiltration with low pressure air) can be considered to ensure that leakage is within allowable limits.

Infiltration Test As per Ontario Provincial Standard Specification (OPSS-410) [12], infiltration tests shall be conducted where the groundwater level at the time of testing is 600 mm or more above the crown of the pipe for the entire length of the test section. The test section is normally between adjacent maintenance holes. A watertight bulkhead is constructed at the upstream end of the test section. All service laterals, stubs, and fittings are plugged or capped to prevent water entering at these locations.

A V-notch weir or other suitable measuring device is installed at the downstream end of the test section. Infiltrating water is allowed to build up behind the weir until the flow through the V-notch has stabilized. The rate of flow is then measured. The rate of flow shall not exceed the maximum allowable infiltration calculated for the test section. Allowable infiltration is calculated as 0.075 l/mm diameter/100 m of pipe sewer/h.

Exfiltration Test **[12]** Exfiltration tests utilizing water or low pressure air shall be conducted on pipe sewers 600 mm in diameter and smaller where the groundwater level is lower than 600 mm above the crown of the pipe or the highest point of the highest service connection included in the test section.

The test section is normally between adjacent maintenance holes. The test section of the pipe sewer shall be isolated by temporarily plugging the downstream end and all incoming pipes of the upstream maintenance hole. All service laterals, stubs, and fittings are plugged or capped to prevent water entering at these locations.

(i) *Testing with Water*

The test section shall be slowly filled with water making sure that all air is removed from the line. A period of 24 h for absorption or expansion shall be allowed before starting the test, except if exfiltration requirements are met by a test carried out during the absorption period.

Water shall be added to the pipeline prior to testing until there is a head in the upstream maintenance hole of 600 mm minimum over the crown of the pipe or at least 600 mm above the existing groundwater level, whichever is greater. The maximum limit of the net internal head on the line is 8 m. In calculating net internal head, allowance for groundwater head, if any, shall be made. The distance from the maintenance hole frame to the surface of the water shall be measured. After allowing the water to stand for 1 h, the distance from the frame to the surface of the water shall again be measured. The leakage shall be calculated using volumes.

The leakage at the end of the test period shall not exceed the maximum allowable calculated for the test section. Allowable leakage is calculated as 0.075 l/mm diameter/100 m of pipe sewer/hour. An allowance of 3.0 l/h/m of head above the invert for each maintenance hole included in the test section shall be made.

(ii) *Testing with Low Pressure Air*

The Contract Administrator may allow or require testing by the use of air where water is not readily available or the differential head in the test section is greater than 8 m or freezing temperatures exist. Air control equipment that includes a shut-off valve, safety valve, pressure-regulating valve, pressure reduction valve, and monitoring pressure gauge with pressure range from 0 to 35 kPa with minimum divisions of 0.5 kPa and accuracy of approximately 0.25 kPa shall be provided.

Tests shall be conducted between two consecutive maintenance holes. The test section shall be plugged at each end. One plug shall be equipped with an air inlet connection to fill the pipe sewer system with air. The test section shall be filled slowly until a constant pressure of 24 kPa is maintained. If the groundwater is above the pipe sewer being tested, the air pressure shall be increased by 3.0 kPa for each 300 mm that the groundwater level is above the invert of the pipe.

Table 13.3 Exfiltration test – low pressure air testing [12]

Nominal pipe size (mm)	Minimum time (Min: Sec)	Length for minimum time (m)	Time for longer length (Sec)
100	1:53	182	0.623
150	2:50	121	1.140
200	3:47	91	2.493
250	4:43	73	3.893
300	5:40	61	5.606
375	7:05	48	8.761
450	8:30	41	12.615
525	9:55	35	17.171
600	11:20	30	22.425
675	12:45	27	28.382
750	14:10	24	35.040
825	15:35	22	42.397
900	17:00	20	50.450

The air pressure shall be stabilized for 5 min and then regulated to maintain at 20.5 kPa plus the allowance for groundwater, if any. After the stabilization period, the time taken for a pressure loss of 3.5 kPa shall be recorded.

The time taken for a pressure drop of 3.5 kPa shall not be less than the times shown in Table 13.3. If the length of the test section is greater than the length for minimum time, the new testing time shall be a product of the length of test section multiplied by the time shown in Table 13.3 for the appropriate size pipe. If the results of an air test are marginal, the Contract Administrator may require the section to be retested using water.

(i) *Closed-Circuit Television (CCTV) Inspection*: Closed-circuit television is used for the inspection on new and existing storm and sanitary sewers, water main, pipe culverts, or other accessible conduits to determine whether the sewer line is clean or clogged. A fixed camera is usually used for pipelines less than 300 mm in diameter; however, for pipelines equal to or greater than 300 mm, a pan and tilt camera shall be used. The rotating camera and light head configuration shall have the capability of panning 360° with pan and tilt capability of providing a full view of the pipe to ensure complete inspection of the mainline pipe and service laterals.

Prior to performing CCTV inspection activities, the Contractor shall thoroughly clean the sewer line(s) and service laterals designated to be televised. Prior to performing the video inspection procedure, water must be introduced into the nearest upstream manhole until observed at the nearest downstream manhole. This will insure that any pipe segments with bellies are easily identified during CCTV inspection. All fog shall be evacuated from the pipeline, and the pipeline shall be kept clear of any fog during the CCTV inspection process.

During the test, the interior of the pipe shall be carefully inspected to determine the location and extent of all deficiencies. Pipe conditions that result in a question of proper installation procedures shall be noted so that these conditions can be reviewed and, if necessary, corrected before actual acceptance of the sewer system. At all service connections, the camera shall be stopped, and the pan and tilt features shall be used to obtain a clear picture. At each service lateral, the camera shall be panned to view up each lateral or point of connection, and deficiencies, if any, should be recorded. The inspections should be recorded in color.

(j) A*sphalt Mix Testing:* On construction sites, the temperature of asphalt mix is primarily checked on arrival. Most of the codes recommend the following temperatures:

Delivery: minimum 140 deg. C
Laying: minimum 130 deg. C
Rolling: minimum 120 deg. C

After the asphalt mix is laid and compacted, it is tested for density, stability, and thickness. Usually one pair of cores is taken from every 1000 m^2 area laid and sent to a laboratory for testing. The mean value of field density of the samples should not be less than 97% of the "job standard mix density." The mix design density (job standard mix) is usually taken as 2.355 g/cm^3 (2.355 × 62.4 lbs/ft^3).

The minimum requirement for stability is 8.0 KN. The cores cut for testing are also examined to find the thickness of the asphalt layer. Materials, which do not meet the required tests, should be removed and replaced.

Nonconformity The implementation of corrective action starts with the detection of nonconformance of material or workmanship quality. Rectification works for materials or processes, which do not conform to the project specifications, should be carried out immediately. Records identifying the defects and steps taken for rectification, repair, and rejection or redone should be recorded.

Appropriate steps should be taken to prevent the recurrence of nonconformity; this may require changing the source of supply, revising the material specification, review of design, etc. Sufficient controls of material and workmanship should be implemented to prevent a recurrence of the problem. When the preventive measures are implemented, their effect should be monitored in order to ensure that desired goals are met.

References and Further Reading

1. FIDIC – International Federation of Consulting Engineers. Article on "Quality In Construction".
2. The FIDIC. (1999). Conditions of contract for construction. Red Book; Author FIDIC.
3. CCDC-2. (2008). Canadian Construction Document Committee – Stipulated price contract.
4. Schexnayder, C. J., & Mayo, R. E. (2004). *Construction management fundamentals* (1st ed.). McGraw-Hill Higher Education.
5. Pyzdek, T. (2000). *Quality engineering handbook.* QA Publishing, LLC Arizona.

6. Surahyo, A. H. (2002). *Concrete construction – Practical problems and solutions*. Karchi: Oxford University Press.
7. OPSS -1350. (1996). Ontario provincial standard specification – Material specification for concrete- Materials and products.
8. CSA (Canadian Standards Association) – A23.1-09/A23.2-09. Concrete materials and methods of concrete construction/Test methods and standard practices for concrete.
9. OPSS 701. (2006). Ontario provincial specification standard – Construction specifications for water-main installation in open cut.
10. American Water Works Association (AWWA)- C651-05- Disinfecting Watermains.
11. Ontario Building Code. (2006). Part-7 plumbing. Building Code Act & Ontario Regulations 350/06.
12. OPSS 410. (2008). Ontario provincial standard specification – construction specification for pipe sewer installation in open cut.

Chapter 14
Construction Safety

14.1 Introduction

Safety on a construction site is one of the most important concerns in the construction industry. Construction sites are one of the most dangerous places to work as large and heavy equipment is used on-site. Construction is an industry where hazards and risks change continuously. Construction site work-related accidents can lead to serious injury or even death.

Numerous accidents have been attributed to the lack of clear responsibilities for various site-related issues. Local labor laws for health and safety in construction are always there, but the administrative and procedural requirements are not always properly addressed within the Contract Documents. All of these accidents can be avoided or minimized if proper health and safety training is provided to the workers.

Effective improvements in construction site safety can be achieved by clearly assigning safety responsibilities in the Contract Documents and through a committed and cooperative relationship between the participating parties. Participating parties must agree in advance, how safety-related problems would be solved on each particular project.

14.2 Contract Requirements

Generally under most contracts, the Contractor is responsible for the safety of all site operations and methods of construction. CCDC-2 [1] specifies under clause 9.4 that the Contractor is solely responsible for construction safety at the job site and shall comply with all applicable rules and regulations of construction health and safety legislation. Hence, the Contractor is most often considered the "Constructor" on-site meaning he will be responsible for the health and safety on the entire project.

A. Surahyo, *Understanding Construction Contracts*,
DOI 10.1007/978-3-319-66685-3_14

Section 1 of the Ontario Occupational Health and Safety Act defines "Constructor" as "a person who undertakes a project for an Owner and includes an Owner who undertakes all or part of a project by himself or by more than one employer" [2]. Here Employer refers to contractors and subcontractors.

The Constructor is the party who remains ultimately responsible for the health and safety of all workers. The Constructor should also ensure that all the employers and workers on the project comply with the Act and its regulations. The intent of the Occupational Health and Safety Act (OHSA) is to have one person with overall authority for health and safety matters on a project. This person is the constructor of the project.

Some of the responsibilities that Constructors have on projects as specified by OHSA under Part I of Regulations are [2]:

1. Before beginning work at a project, each Constructor and Employer engaged in construction shall complete an approved registration form.
2. An Employer of five or more workers on a project shall appoint a supervisor for the workers and the supervisor shall supervise the work at all times either personally or by having an assistant, who is a competent person.
3. A Constructor shall establish written procedures to be followed in the event of an emergency and shall ensure that the procedures are followed at the project.
4. The Constructor shall ensure that the emergency procedures are posted in a conspicuous place at the project.
5. The Constructor shall ensure that every worker at the project has ready access to a telephone, two-way radio, or other system of two-way communication in the event of an emergency.

However, as per clause 3.2 of CCDC-2 [1], if an Owner awards a separate contract or other parts of the project to a separate Contractor or firm or if work is performed by the Owner's own forces, then the Owner has to assume overall responsibility for compliance with the applicable health and construction safety legislation at the place of the work.

Additionally as per clause 9.2 of CCDC-2 [1], the Owner will also be responsible for control and management of toxic and hazardous substances with reference to existing conditions. The Owner shall take such steps to determine whether any such material is present at site of work and provide related information to the Consultant and Contractor in writing.

The FIDIC [3] Sub-clause 4.8 provides that the Contractor is responsible for the safety of persons on-site and needs to provide facilities, such as, lights, guards, fencing, warning signs, etc., so as to protect the works and the public from injury or damage. As per Sub-clause 4.18, the Contractor should also take all reasonable steps to protect the environment on and around the site to avoid damage or nuisance to persons or public property or others resulting from pollution, noise, or other causes arising as a result of his methods of operation. Additionally, the Contractor has to make sure that emissions, surface discharges, and effluent resulting from his operations shall not exceed the values indicated in the specifications or as prescribed by the applicable local laws.

As per Sub-clause 6.7, FIDIC [3] further specifies that the Contractor is also required to appoint a qualified safety officer on-site dealing only with issues of safety and protection against accidents of all staff and labor. The Contractor has to ensure in collaboration with local health authorities, that medical staff, first aid facilities, and suitable ambulance services are available and arrangements are made for the prevention of epidemics and for all necessary welfare and hygiene requirements.

The FIDIC [3] Sub-clause 2.3 also provides that the Employer is responsible in respect of work he will carry out with his own workmen or with other contractors on the site.

In addition to the contractual obligations of the Contractor and the Owner, safety largely depends upon training, vigilance, and cooperation during the construction process. Workers should be constantly alert to the possibilities of accidents and avoid taking unnecessary risks. Good project managers and contract supervisors should make sure that the job is done right the first time and that no major accidents occur on the project.

14.3 Common Safety and Health Issues and Guidelines

Injuries at work remain a major cause of death and disability, whereas the major causes of injuries and death are falls and falling objects. As per data compiled by OSHA (Occupational Safety and Health Administration, U.S. Department of Labor), about 2.3 million construction workers frequently work on scaffolds. Protecting these workers from scaffold-related accidents would prevent an estimated 4500 injuries and 50 fatalities each year [4]. Also as per Ministry of Labour, Ontario, falls are the number one cause of critical injuries and deaths of workers at construction sites in Ontario [5].

In addition to falls and falling objects, excavated trench collapses, inadequate operation of machineries, and exposures to chemicals, dusts, vibration, and noise are also hazardous to health. People who work with vibrating machinery may develop occupational hearing loss. People who work with asbestos, tar or tar derivatives, certain chemicals, or radioactive compounds are at increased risk of various types of cancer.

The health of these workers needs to be monitored regularly. Today many types of occupational hazards are recognized and publicized, and employers are legally required to take steps to minimize injuries. In general, the risk of injury or death could be reduced by closely following the safety standards. Workers need to be trained in "Work Place Hazardous Materials Information System" (WHIMS) which is Canada wide training system for the people who work with hazardous materials on the job providing information to protect their health and safety. Some of the protection guidelines are addressed hereunder:

14.3.1 Personal Protective Equipment (PPE)

Personal protective equipment describes types of protective clothing and equipment that workers need to work safely. PPE should fit properly, and workers should know how to use them. Workers must understand and follow all safety rules at all times and use such personal protective equipment as are necessary to protect them against any possible site hazards. Everyone on construction site must:

1. Wear safety shoes that extend above the ankle or safety boots.
2. Wear a bright orange or fluorescent yellow vest for high visibility.
3. Wear a certified safety-approved helmet or headwear.
4. Wear gloves, a mask, and safety goggles when handling dangerous chemicals.
5. Wear a mask to protect the lungs from inhaled fumes or particles when working in a dusty environment.
6. Wear disposable foam ear plugs in the vicinity of high intensity noise sources.
7. Avoid to lift, carry, or move any load so heavy as to be likely to cause injury.
8. Wear safety glasses when there is a risk of eye injury.

14.3.2 Protection from Falls and Falling Objects

Falls and falling objects can result from unstable working surfaces, ladders that are not safely positioned, or missing protective devices, such as guardrails. Workers are also subject to falls or to the dangers of falling objects if sides and edges, floor holes, and wall openings are not protected.

In general, if workers are at a height of more than 1.2 m and the work area is used as a path for a wheelbarrow or similar equipment or otherwise more than 3-m height, the worker must be protected by a fall protection barrier. It is better to use fall prevention systems, such as guardrails, than fall protection systems, such as safety nets or fall arrest devices.

However, sometimes it is not possible to install a guardrail system like on a roof of a building or house; a worker then must use one of the following protection systems as recommended by Ontario Occupational Health and Safety Act (OHSA) [2]:

1. A travel restraint system consisting of a full body harness or safety belt attached by a lifeline or lanyard to a fixed support, such as a permanent anchor system (Section-26.4)
2. A fall-restricting system attached to an independent fixed support, such as a permanent anchor system – the system be designed to limit a worker's free fall to 0.6 m (Section-26.5)
3. A safety net that needs to be designed, tested, and installed in accordance with the American National Standards Institute (ANSI) Standard 10.11.1989 (Section-26.8)

Additional safety measures required are:

1. To make sure that equipment like ladders and scaffolding are adequately and firmly secured.
2. To ensure that scaffolding working platforms should consist of good quality timber boards at least 460-mm wide and well supported.
3. Every scaffold fitted with wheels shall have a locking device for the wheels which shall be properly secured when scaffold is used.
4. Every working platform from which a person is liable to fall a distance of more than 2 m shall be provided with guardrails to a height of at least 1 m above the platform. The guardrail system shall have a top rail, an intermediate rail, and a toe board of at least 15-cm height.
5. All scaffold uprights must be braced adequately to prevent lateral movement.
6. Every scaffold shall be designed and constructed to support or resist two times the maximum load or force to which it is likely to be subjected, without exceeding the allowable unit stresses for the materials of which it is made and four times the maximum load or force to which it is likely to be subjected without overturning (Section-126 of Regulations) [2].
7. No scaffold shall be overloaded, and any load placed thereon shall be evenly distributed.
8. Scaffold materials, tools, and other waste materials shall not be thrown, tipped, or shot down from a height where they are liable to cause injury but shall be properly lowered except where a properly constructed chute is in use.
9. Scaffolds must be at least 10 feet (about 3 m) from electric power line at all times.
10. Avoid using ladders with metallic components near electric work and overhead power lines.
11. Every ladder used as an access or part of a scaffold or as a working place shall be properly secured to prevent slipping or sliding at the head or foot and undue swaying.
12. Every ladder should project above any landing or working point for a distance of at least 900 mm.

14.3.3 Protection from Machinery Hazard

People can be struck and injured by moving parts of machinery, sharp edges can cause cuts and severe injuries, people can be crushed, and injuries can also occur when machines are used by inexperienced operators. General safety precautions are:

1. On using any type of machinery, such as cranes, forklifts, trucks, tractors, and bulldozers, make sure that they are being driven by a competent properly trained and licensed operator.

2. While operating machinery, ensure that the required guards and shields are in place. Machine guards prevent exposure to moving, rotating, and electrically charged or hot parts.
3. Make sure that operators of dangerous machinery do not become overtired.
4. Workers exposed to public vehicular traffic should wear warning vests.
5. No employee shall be permitted underneath loads handled by lifting or digging equipment. Employees shall be required to stand away from any vehicle being loaded or unloaded to avoid being struck by any spillage or falling materials.
6. Before using a crane, its controls should be checked for proper operation along with its wire rope, chains, and hook.
7. Ensure that the load does not exceed the crane's rated load-carrying capacity.
8. Ensure tower crane foundations; shoring and bracing are designed by a professional Engineer.
9. Do not move a load over workers and barricade accessible areas within the crane's swing radius.
10. Adequate clearance shall be maintained between moving and rotating structures of the crane and fixed objects to allow the passage of employees without harm.
11. Each employee required to perform duties on the horizontal boom of hammerhead tower cranes shall be protected against falling by guardrails or by a personal fall arrest system.
12. Ensure that the maximum acceptable load and the last test results are posted on the crane.
13. Watch for overhead electric distribution and transmission lines and maintain a safe working clearance of at least 10 feet (about 3 m) from energized electrical lines.

14.3.4 Protection from Trenching Hazard

A trench is defined as an excavation in which the depth exceeds the width. A significant number of deaths and injuries are directly related to trenching for sewer and water main works. Most cave–ins occur on small, short-duration jobs like water, gas, electrical, and sewer connections. Trenching fatalities are mainly caused by cave–ins. Death occurs by suffocation or crushing when a worker is buried by falling soil. The other hazards resulting in injuries could be:

1. Slips and falls into trenches or excavations
2. Tripping over equipment
3. Excavated material or other objects falling on workers
4. Exposure to underground electrical cables
5. Unstable adjacent structures
6. Poorly placed excavated materials
7. Exposure to toxic, irritating, or flammable gases

Soil properties often vary widely throughout the length of the trench. Some of the main factors affecting trench stability are:

1. *Soil type*: Type of soil determines the strength and stability of trench walls. Ontario Occupational Health and Safety Act (OHSA) [2] under Section-226 of the Regulations classify soil as types 1, 2, 3, and 4. Hard and very dense soil is considered as Type1, very stiff and dense as Type 2 soil, stiff to firm as Type 3, and soft to very soft and very loose as Type 4 soil.

   - Based on soil classifications, there are three methods of protecting workers against trench cave–ins which are sloping, trench boxes, and shoring. It is always recommended that no one enters an excavated trench deeper than 1.2 m, unless it is properly sloped, shored, or properly supported by a trench box.
   - With respect to sloping, OHSA [2] under Section-234 of Regulations recommends that for types 1 and 2 soils, trench walls above depth of 1.2 m should be trimmed at 1:1 slope. For Type 3 soil, overall trench walls are to be cut at 1:1 slope. For Type 4 soil, OHSA [2] recommends 3:1 slope (three horizontal, one vertical).
   - Trench boxes are used mainly to structurally uphold trench walls so as the workers can work in a protected portion of the trench. As long as workers are in the trench, they have to remain inside the box and leave only when the box is being moved. A ladder must be set up in the trench box at all times. It is also required that the space between the trench walls and the box should be restricted to the minimum clearance required for the forward progression of the box.
   - Shoring is a system which is provided to shore up and support trench walls to prevent movement of soil. Timber and hydraulic systems are the most commonly used supports to shore up walls. Both types must be designed by a professional Engineer. Hydraulic shoring means prefabricated systems in aluminum or steel. OHSA [2] under Section-237 of the Regulations suggest that no hydraulic or prefabricated system should be used in Type 4 soil. The space between the hydraulic support system and the trench walls should be minimized leaving only the clearance required to allow the support system to be moved forward. It is always advisable that shoring should be installed as excavation proceeds and from top down.

2. *Trench wall stability*: Equipment vibrations or stock piling excavated material adjacent to trench greatly affects the stability of trench walls. Always keep heavy equipment, excavated soil, or construction material away from the edges of the trench. At least 1 m of each wall's upper edge must be kept clear of equipment, excavated soil, or construction material. No vehicle or machinery should be operated within 1–2 m of an excavated trench.

   The walls of an excavation should always be stripped of any loose rock or other material that may slide, roll, or fall on a worker to avoid any injury. A cleared work space of at least 450 mm between the wall of an excavation and any formwork, masonry, or similar wall should be maintained (OHSA Section-232&231) [2].

3. *Fall protection*: OHSA under Section-233 of the Regulations suggests that a barrier is to be provided, at least 1.1-m high, at the top of a trench, hole, or shaft if the excavation does not meet regulatory slope requirements, and is more than 2.4-m deep [2]. Unless walls of a trench are sound and made of solid rock, never enter a trench deeper than 1.2 m unless it is properly sloped or protected. No one should work alone in a trench or under suspended or raised loads and materials.

4. *Underground services*: OHSA under Section-228 of the Regulations recommends that Employers must ensure all gas, electrical, and other services are located and marked in or near the area to be excavated [2]. Tests must be conducted for atmospheric hazards such as low oxygen, hazardous fumes, and toxic gases prior to excavation. If a service poses a hazard, it must be shut off and disconnected before the excavation activity begins. If a potentially hazardous service cannot be disconnected, the service Owner must be asked to supervise its uncovering during the excavation.

14.3.5 Protection from Confined Space Hazards

Confined space is defined as a fully or partially enclosed space that is not designed and constructed for continuous human occupancy, which may contain a hazardous atmosphere because of its construction, location, contents, or work that is done in it. Common examples are water and sewage tanks, manholes, trenches, septic tanks, digesters, and cable vaults.

Atmosphere is considered hazards when [6]:

1. An oxygen content in the atmosphere is recorded at less than 19.5% or more than 23% by volume.
2. The lower explosive limits (LEL) is more than 10%. Work can proceed when LEL is 0%, and cold work is allowed if LEL is 0–10%.
3. Carbon monoxide (CO) is more than 25 ppm.
4. Hydrogen sulfide (H2S) is more than 10 ppm.

Before any worker enters a confined space, an adequate written plan should be developed including procedures for controlling the possible hazards. Each worker who enters a confined space must wear adequate protective clothing and personal equipment and devices, in accordance with the relevant plan.

No person shall enter a confined space until the atmosphere of the confined space has been tested by using a calibrated gas detector. If the atmosphere testing indicates that a hazard exists, entry must not be allowed until the space has been adequately ventilated, and subsequent tests indicate that the air is safe to breathe.

If the atmosphere testing indicates no hazard exists, a worker may enter the confined space provided that the atmosphere in the confined space is continuously monitored, and the worker wears a full body harness and to be tied to a rescue line. A worker/attendant must be on top in case of an emergency. The attendant needs to monitor the safety of the worker (s) inside the confined space through adequate means of communication and atmospheric testing and be immediately available to provide emergency assistance.

14.3.6 Protection from Electrical Hazard

Electrocutions are also one of the major hazards on construction sites. In general, there are two ways to be injured by electricity. One is by electric shock and the other is by arc flash. Electric shock is the passing of electric current through the body which can cause loss of balance, increasing fall risk. Higher voltages/amperages may result in severe internal and external burns and might kill a person on contact.

Electricity flows through conductors, which includes metals, water, the Earth, and the human body. Electricity must have a complete circuit or path to flow. When electrical tools are working properly, a complete circuit is maintained between the tool and the energy source. However, if the tool is damaged, a person may come in contact with the electricity and can become a path for the current. In this event the person will be shocked.

In addition, an arc flash is a release of energy caused by an electric arc. The release of energy is often referred to as an arc flash. The flash causes an explosive expansion of air and metal. The blast produces a dangerous pressure wave resulting in blast injuries, lung injuries, ruptured eardrums, severe burns, etc. Arc flash injuries can also result in death.

The following is a list of a common electrical hazards (including necessary precautions to mitigate the hazards) found on construction sites:

1. Improper grounding: Electrical equipment must be properly grounded. Grounding reduces the risk of being shocked or electrocuted.
2. Exposed electrical parts: Exposed wires or terminals are hazardous. These conditions must be reported to a supervisor.
3. Inadequate wiring: Ensure that power tools are being used with a properly rated extension cord. Extension cords should be protected during use to prevent damage, and damaged tools or extension cords should not be used. Extension cords should be plugged into Class A ground fault circuit interrupters (GFCIs).
4. Overloaded circuits: Overloaded circuits can cause fires. Proper circuit breakers should always be used. Never overload an outlet. Always avoid using tools in wet locations. Water increases the risk of electric shock.
5. Overhead power lines: Overhead power lines are very dangerous. At least a distance of 3 m (10 ft) between tools and equipment and overhead power lines should be maintained.

6. Protective tools and equipment: Workers exposed to an electrical hazard must use mats, rubber gloves, shields, flame resistant clothing, and any other protective equipment required to protect themselves from electric shock and burn.
7. Lockout and tagging: An important aspect of electrical work involves isolating electrical energy. Lockout means disengaging all energy sources from equipment, safely releasing stored energy or materials, and securing the control devices in the "off" position with an approved lock, thereby rendering it incapable of operation, release, or movement. Lockout tag means an approved tag provided by the supervisor in charge that allows work to be performed on the equipment or process. Workers must ensure electricity is off and "locked-out" before work is performed.

For further information on health and safety issues, readers are advised to review guidelines provided online by several organizations, such as Occupational Safety and Health Administration (OSHA) USA, Occupational Health and Safety Act (OHSA) Canada, Health and Safety Executive (HSE) UK, etc.

References and Further Reading

1. CCDC-2. (2008). Canadian Construction Document Committee- Stipulated price contract.
2. Ontario Occupational Health & Safety Act (OHSA- R.S.O 1990) and Ontario Regulations 213/91 for Construction Projects- Queen's Printer for Ontario; www.ontario.ca/laws
3. FIDIC. (1999). *Conditions of contract for construction*. Red Book: Author FIDIC (International Federation of Consulting Engineers).
4. OSHA. (2005). 3252- 05N – Occupational safety & health administration – Pocket guide worker safety series – Construction. Department of Labor US.
5. Ontario Ministry of Labour (Last Reviewed 2014). Article on "Preventing Falls on Construction Projects".
6. Confined Space Entry Course. (2009). Attended by the Author through city of Barrie Ontario-Course Offered by "TEAM-1 Academy".

Chapter 15
Construction Claims

15.1 Introduction

Changes and claims are a common element of today's construction industry, particularly where the project is complex and may result in millions of dollars of extra payment. Construction changes were also discussed in a previous chapter; here mainly construction claims or unresolved changes will be addressed. A claim is an unresolved change and if a change is agreed and processed, it will not turn to a claim. It is therefore beneficial to both the Owner and the Contractor to resolve the changes promptly, fairly, and sensibly to avoid reaching the claim situation. However, the major role of all parties to the contract should be to avoid problems or to minimize their impact when they do arise.

15.2 Construction Claims

15.2.1 Definition of a Claim

A claim is a written statement by one party seeking, as a matter of right, additional time and/or money for acts or omissions by other party during performance of the contract with respect to the terms of the contract. In fact, claims are an integral part of the construction process, and good claim and administration principles are equally important as managing quality and safety, etc. on construction sites. The provisions are outlined in most of the standard forms of contract, under which the Contractor can claim against the Owner for any losses suffered if the progress of work is affected due to certain specified causes.

Early identification and notification are the most important factors of the entire claim process. Prompt identification enables the Employer to verify, confirm, and

A. Surahyo, *Understanding Construction Contracts*,
DOI 10.1007/978-3-319-66685-3_15

possibly remedy the situation at the earliest possible. With early identification, the Contractor becomes in a position to study the problems, analyze different proposals and notification options, so as to prepare the claim properly without any trouble.

15.3 Types/Causes of Claims

There are many causes of claims; however, some of the consistent causes of claims are summarized here:

(a) *Delays & Accelerations*: Delay and accelerations are an extremely common source of construction claims.

(i) ***Delay***: Delay refers to the situation in which the activity of one party is affected by the inactivity, inability, or constraints of another party. Progress delays occur in many construction projects. Examples could be:

1. Delay in handing over site to the Contractor
2. Delay in approvals and decisions
3. Delay due to Owner's nominated subcontractors or by the General Contractor's subcontractors
4. Delay or defects in items supplied by the Owner
5. Delay in delivery and supply of materials
6. Delay in progress of work by the Contractor
7. Failure to adequately schedule and coordinate the work
8. Strikes, adverse weather, Contractor errors, variations and Owner-directed suspension, and so on.

Progress delays are of two types: excusable delays and non-excusable delays.

Excusable delays Excusable are delays to the completion of works arising from conditions beyond the control (and without the fault or negligence) of the parties, e.g., acts of God, fires, floods, strikes, severe weather, etc. An excusable delay is one that justifies an extension of time.

Non-excusable delays Non-excusable delays do not justify an extension of time. Non-excusable delays are those that are generally within the control of the Contractor or have been contractually assigned to the Contractor. The examples could be delay in submission of shop drawings, lack of sufficient resources, subcontractor delays, etc.

Excusable delays are further classified into compensable and non-compensable delays.

Compensable delays Compensable delays occur when one party fails, without excuse to timely perform a contractual duty, which causes a delay in the other party's performance. The examples could be poor planning and management of Contractor's site staff, lack of resources, delay in handing over site by the Owner, design errors, etc. Compensable delays authorize both an extension of completion time as well as collection of damages caused by the delay. Compensable delay may result in the contractor's termination for default or the assessment of liquidated damages.

According to CCDC 2 [1] clause 6.5, if the Contractor is delayed in the performance of the work because of the Owner, Consultant, or anyone employed or engaged by them directly or indirectly, or by a court or other public authority, without any fault of the Contractor, then the Contractor shall be entitled for reasonable time extension and costs incurred by the Contractor as the result of such delay.

Non-compensable delays There are two forms of non-compensable delay impacts [2]:

1. An excusable delay may be non-compensable if the project delay is beyond the control of both parties. Inclement weather and acts of God may entitle the Contractor to a time extension but not monetary compensation.
2. An excusable delay is non-compensable if the conditions could have been avoided by the Contractor. For example, if the project was delayed by the Contractor's actions or inactions, the Contractor would not be entitled to time or money for the delay.

(ii) **Acceleration**

Acceleration is speeding up the progress by the Contractor to achieve early completion or to make up for time lost due to delays unless a time extension is granted. Acceleration can be classified as (1) voluntary, (2) directed, or (3) constructive.

Under voluntary acceleration, the Contractor at his own initiative decides to speed up the progress either to cover his own delays or to complete the project earlier than planned.

Directed acceleration occurs where an order to accelerate performance is given by the Owner. The contractor normally is entitled to compensation for any extra expense caused by an acceleration order.

Constructive acceleration occurs in the absence of an Owner directed acceleration or where the Contractor has not actually been expressly directed to accelerate. This happens when the Owner has refused a valid request for time extensions or threatened other action.

As a result, if the Contractor accelerates its performance (to avoid liquidated damages or other loss or risk of loss) and incurs additional expense in order to overcome the delay and achieve the original completion date, then the Contractor's performance has been "constructively" accelerated, even though no direct acceleration order has been given. The Contractor normally is entitled to compensation for constructive acceleration.

It is important that the Contractor advises the Owner/Contract Administrator that any efforts undertaken by the Contractor to accelerate performance are not voluntary, especially when there is the perception of acceleration but an instruction has not been received.

The principle of constructive acceleration has been applied in the United States and Canada but may vary in other countries. Canadian cases such as *Morrison-Knudsen Co. v. British Columbia Hydro & Power Authority* [3] and *W.A. Stephenson Construction (Western) Ltd. v. Metro Canada Ltd.* [4] can be referenced in this context.

(b) *Directed and Constructive Changes*: Directed change refers to the change in scope of work, i.e., addition or deletion of works or the use of different methods, materials, or design. It is a directive by an Owner or his authorized agent through a written change order to make a change to the contract.

Constructive change, also known as "force account work," is not the same as directive change, as no written change order is issued for the change with respect to work required. An oral or written communication by an Owner or his authorized Agent is considered having the same effect as a written change order.

Whenever a change during work occurs and the change is not the responsibility of the Contractor, the Contractor should immediately give written notice to an Owner that a change has been made. Contractors should not wait for final calculation of cost or delay. Sometimes, on a job with many changes, the Contractor at the end of job is in a loss position and submits claims for the total impact of the changes. Such claims are usually denied because the impacts and delays are not traceable to specific cause [5].

(c) *Concurrent Delays*: A concurrent delay occurs when two or more independent causes of delay overlap in time [6]. For example, if a Contractor-caused delay is 2 weeks and an Owner-caused delay is 1 week and they do not overlap, then the Contractor would be entitled for only 1 week time extension for the nonconcurrent Owner-caused delay, whereas if they overlap, then the Contractor would not be entitled to an extension of time because of concurrent delay. In another example, if the Contractor-caused delay is 1 week and Owner-caused delay is 2 weeks, and they overlap, then the Contractor would be entitled to get 1 week extension for the non-overlapping week.

(d) *Cardinal Changes*: A cardinal change made by the Owner is one where the purpose of the original contract has been frustrated or made impracticable or impossible by the magnitude of the requested change. A cardinal change is mostly viewed as a breach of contract on the part of Owner, and the Contractor may either refuse to perform the work or may elect to complete the work under protest and recover reasonable cost for performing the changed work.

(e) *Differing or Changed Site Conditions*: This situation refers to the changes encountered during work performance which were not visible and not known at the time of bidding. Differing site conditions are addressed under FIDIC [7] Sub-clause 4.12 as "unforeseeable physical conditions" and under CCDC2 [1] clause 6.4 as "concealed or unknown conditions." Unforeseeable physical conditions are also the most common source of claims and disputes in construction projects. According to FIDIC [7] Sub-clause 4.12, physical conditions mean natural physical conditions (such as change in soil characteristics, encountering rock, etc.) and man-made (such as unidentified utilities, buried structures) and other physical obstructions and pollutants including subsurface and hydrological conditions but excluding climatic conditions. The most common problems are changes in subsurface soil conditions and unidentified subsurface utility lines.

According to FIDIC [7] Sub-clause 4.10, the Employer has to make available to the Contractor for his information all available details of subsurface and hydrological conditions at the site including environmental aspects. The

Contractor shall be responsible for interpreting all such data and making his own inquiries in order to satisfy himself that the tender is adequate.

However, if the site conditions actually are concealed from the Contractor, then the concealment of the site conditions may give rise to a tort claim for fraud and misrepresentation.

If a construction contract represents that certain conditions exists on the site and in fact the actual conditions on the site differ from those represented in the contract drawings and specifications or the Contractor encounters adverse physical conditions which he considers to have been unforeseeable, then a contract-based claim arises.

CCDC2 [1] under clause 6.4 provides that if the Owner or the Contractor discovers that subsurface or concealed physical conditions, other than conditions due to weather, differ materially from those provided in the Contract Documents, then the observing party shall give notice in writing to the other party within 5 working days without disturbing the discovered conditions. The Consultant shall investigate such conditions which may cause an increase or decrease in the Contractor's cost or time to perform the work. The Consultant with the approval of the Owner will issue appropriate instructions for a change in the work.

As mentioned above, it may be noted that claims with respect to differing site conditions must be for physical conditions. Claims are not entertained under nonphysical conditions such as labor disputes, governmental, political, and economic, or weather conditions.

Most standard Contract Documents, identify two types of differing or changed site conditions as Type-1 and Type-2.

Type-1 Changed Conditions Type-1 condition refers to subsurface or concealed (hidden) physical conditions, differing materially from those provided in the Contract Documents. Type-1 conditions should not be considered as subsurface conditions only, but they also cover conditions which are at or above surface that are concealed or hidden, for example, presence of plumbing in walls, which is not shown or identified in the contract documents at the time of bidding. Some other examples of Type-1 conditions include:

1. Presence of subsurface water or encountering rock during subsoil excavation, not indicated in the Contract Documents.
2. Presence of permafrost, not indicated in the Contract Documents. National Geographic describes permafrost as a permanently frozen layer below the earth's surface, consisting of soil, gravel and sand usually bound together by ice.

Type-2 Changed Conditions Type-2 condition is described by most Contract Documents as unknown physical conditions of an unusual nature, differing materially from those ordinarily encountered and generally recognized as inhering in the work [8]. By definition, a Type-2 condition is not described in the Contract Documents. It must be an unknown physical condition of an unusual nature [9]. Example could be that during excavation a water table was encountered to be much higher than reasonably could have been anticipated in such type of works, within surrounding areas.

(f) *Inclement weather*: CCDC-2 [1] defines this as abnormally adverse weather (Sub-clause 6.5.3), whereas FIDIC [7] addresses this condition as exceptionally adverse conditions (Sub-clause 8.4). Both standard forms in fact refer to extremely adverse weather conditions rather than merely bad weather. Primarily, inclement weather conditions are considered as an excusable delay for which extension of time is allowed without monetary compensation. However, the Contractor has to prove that the climate conditions were abnormally adverse and actually delayed the completion of the project.

(g) *Force Majeure*: Force majeure refers to the events beyond the control of parties. Like inclement weather conditions, this is considered by most of the standard forms as an excusable delay for which extension of time is allowed only. Force majeure is also explained in Chaps. 10 and 16.

(h) *Permits*: Acquisition of permits and approvals from outside agencies.

(i) *Termination*: Contract suspension or termination. It is further explained in Chap. 16

(j) *Breach of contract*: This is also explained in Chap. 16.

(k) *Implied warranty*: Duty to perform work in a skillful, careful, and diligent and workman-like manner. Breach of implied warranty is considered a tort, and the most common implied warranty claims are discussed in Chap. 18.

(l) *Defective and deficient contract documents*: Contract Documents that contain errors and omissions, code violations, or a lack of inter-discipline coordination. The Owner would be considered responsible for delays caused by defective plans and specifications, including the failure to make those revisions in a timely fashion.

(m) *Hiding Facts*: Conditions/facts withhold knowingly by the Owner that will adversely affect the Contractor's performance. Such type of claim is known as "superior knowledge" by the court system.

(n) *Disruption*: Disruptions are impacts which delay specific activities but not the overall project completion date. Disruptions are situations which lead to loss of output and production or performance of the work in a less efficient manner. Sometimes the work takes more time to complete, using the original/same resources, or the work takes the same time but requires increased resources. Lost productivity claims are based upon a variety of factors. Most of these factors result in a disruption which creates problems in production, e.g., changes in sequence of work, disruption to planned continuity of work, etc.

15.4 Dealing with Claims

Identification of claims must be followed by notification. For any type of claim situation described above, the affected party must notify the other party in a timely manner as soon as claims situation arises. CCDC-2 [1] clause 6.6 specifies that if either the Contractor or the Owner intends to make a claim for an increase or credit to the contract price, respectively, that party shall give timely notice in writing of intent to claim to the other party and to the Consultant. Timely notification allows both parties to verify conditions, to assemble facts, and to resolve disputes while the issues are fresh in their minds.

When a Contractor encounters any claim situation, such as differing site conditions or any error or omission on drawings, he should stop that particular activity until a satisfactory response is obtained from the Owner. If the Contractor continues work with the affected activity, that may result in a Contractor's obligation to correct the defective work at his own cost. The Contractor, while submitting notice, should always stress for early resolution of a dispute and should request acknowledgement. Delay in notifying the concerned parties will prolong the resolution of the claim issue.

In addition to notification, proper documentation of all facts related to the claim situation plays a very important role in resolving the dispute. Proper records should be maintained throughout the life of the project, so as to support a claim in a suitable manner. Sometimes it happens that claims, which might be valid, are rejected because of the following:

(a) Timely notice was not submitted by the affected party.
(b) The claim was submitted late.
(c) Contract procedures were ignored or not followed properly.
(d) Proper records were not available.
(e) Inadequate submittal to support claim quantification.
(f) Inadequate proof of claim entitlement under the contract.

Most standard forms of contract stipulate certain procedural steps that must be followed if a claim is to succeed. FIDIC [7] under Sub-clause 20.1 provides a disciplined way of dealing with claims for any extension of time or additional payment to avoid prolonged disputes, which requires the Contractor:

1. Notify his intention to claim for additional payment to the Engineer and Employer within 28 days after the event giving rise to the claim has first arisen.
2. Keep contemporary records.
3. Submit the claim to the Engineer within 42 days after the first notice, along with detailed particulars of the amount claimed and the reasons of claim.

A Contractor, who fails to meet the procedure laid down in the conditions, will limit his entitlement to payment, because without the supply of contemporary record evidence, the Contractor cannot put the claim in a better position. It is therefore necessary that the Contractor should submit a properly documented and verifiable claim containing:

(a) A brief summary of the claim
(b) The contract clauses, on which the claim is based, i.e., entitlement
(c) Reasons and instructions giving rise to the claim
(d) Valuation and justification of the claim, supported by the proper documentation
(e) Details of any delay and time extension applicable

As soon as the Consultant receives the Contractor's "notice of claim," he has to decide whether or not the Contractor's claim is well founded. The Consultant may request the Contractor provide additional information that is reasonably necessary. The Consultant may need to take such steps as to:

(a) Evaluate and investigate the reasons for the claim submitted.
(b) Modify the design, if required, to meet the changed situation.
(c) Take any corrective action that may be required to mitigate the problem.
(d) Inform and discuss the issue with the Owner for agreement.

As per CCDC-2 [1] Sub-clause 6.6.5, the Consultant will provide notice in writing to both parties about his findings on the claim within 30 working days after receipt of the claim by the Consultant, or within such other time period agreed by the parties.

When the Consultant has completed due consultation with the Owner and agrees in principal with the Contractor's claim, the Contractor shall be entitled to get that amount included in any interim payment certified by the Consultant. However, if such finding is not acceptable to either party, then as per CCDC-2 [1] Sub-clause 6.6.6, the claim shall be settled in accordance with the "Dispute Resolution" clause under Part 8 of the General Conditions.

15.5 Maintaining Claim Records

Records and documentation are the most important part of claim administration as it plays a significant role in the successful settlement of contract claims. Maintaining adequate documentation is important to both the Owner and the Contractor. The daily events and details of the work must be documented to substantiate claims and prove damages.

FIDIC [7] Sub-clause 20.1 requires that the Contractor should submit contemporary records to substantiate the claim. A contemporary record is a timely record of an event. If a document is created much later than a few weeks following the event, the document will not be considered "contemporary." If it is noticed that some records are missing a few days after the event, or if there is only time to make a rough contemporaneous note, it is not too late to remedy the situation by supplementing those records a few days or even weeks later.

Contemporaneous project records also serves as an important evidence in courts. The records must be prepared to show that the records were "business records" kept in the ordinary course of business and were relied upon in the operation of the business. If these facts cannot be proven, then the records are of limited value as evidence in a construction trial.

15.5.1 Types of Records Required

There are many types of records that should be maintained during a construction contract. The following pre-construction and post-construction list of documents will be of great help in resolving a construction claim:

1. *Owner's bid package and contractor's bid preparation documents*: These records provide a good evidence as to what was intended by a construction contract. It is particularly useful for the Contractor to know which job or units were above or below originally estimated quantities.
2. *Construction specifications*: Construction specifications, including a complete set of all revisions to the specifications, should be retained.
3. *Construction schedule*: Construction schedule defines how the job will be performed indicating the timing of the major activities of the project, providing sufficient details of the critical activities and their interrelationship to meet the stipulated completion time. In general, the delay must affect the critical path of the work to be compensable. If the delay absorbs the float in the schedule, then it is not compensable. The Contractor usually monitors the actual progress of the work and compares with the base schedule on a regular basis by highlighting on the bar chart as the work continues. Schedules are a great help in analyzing any delay encountered.
4. *Tender drawings, construction drawings, revised drawings, and shop drawings*: Generally, Contractors treat construction drawings very casually which is mainly the record of design changes or errors or which may provide crucial support for a Contractor's claim of extra work. The construction drawings stamped with the date of receipt and kept in chronological order is extremely helpful in monitoring the design development of the project. Dates on revised drawings can establish delay in approval which may result in construction delays.

   Shop drawings also play important role in the claims. Often, the Contractor is required by contract to describe exactly how the work is to be installed. The approval stamp on the drawings by the Consultant can provide crucial evidence that the Contractor's means and methods of construction were known to the Owner's design professionals.
5. *Project correspondence including minutes of meetings*: Every Project Manager or Contract Administrator recognizes the importance of project correspondence, transmittals, notices, and record of meetings. These records provide vital evidence of what happened on the job and when and what was discussed and agreed. It is very helpful to stamp all mail or transmittals with the date the document was received. Often, correspondence is actually received many days after the date on the letter of transmittal, and the only way to prove this later is with a date stamp showing when the document actually was received.
6. *Daily record of labors and plants, concrete pours, material delivery, time record*: These documents are critical job records as they help with calculating any claim, particularly on a time and material basis. Material records are not only used to verify what material was delivered to the site and when but also to verify escalation costs. They can also be reused to establish proof that the Contractor did not cause delays due to failure to have material on time. Daily labor, plant, and time records help in establishing the number of hours worked, when two or more items of work were performed and which machinery used on any given day.
7. *Work progress reports* including cash flows, progress photographs, inspection reports, and test reports.

8. *Measurement record*, interim valuations, change order logs, request for information (RFI) logs, and payment certificates.
9. *Employee payroll records*.
10. *Record of payment* made to suppliers and subcontractors.
11. *Site survey* records, like existing ground levels, etc.
12. *Project diary* indicating record of weather, visitors, deliveries, accidents, and any notable event that may cause delay or affect productivity.

15.6 Pricing of Claims

There are several methods of pricing construction claims. Each of these methods may be acceptable, depending upon the facts and circumstances of each case. However, the preferred methods for pricing construction claims include calculation of actual cost (direct cost), impact costs and overheads (indirect cost), and profits.

A second method of pricing construction claims is the "total cost method." This method includes the actual total cost less the base tendered cost and less the cost of variations/change orders already paid.

The second method, being simple, is less popular for computing adjustments because it becomes difficult to prove the Employer's liability for the extra costs and to prove whether the Contractor was working inefficiently or his tender was underestimated. For these reasons, the total cost approach should only be used where no other method of pricing the claim is available.

15.6.1 Direct and Indirect Costs

1. *Actual Cost or Direct Cost*

If any claim involves carrying out changed work or extra/additional work on site, then the actual cost of the labor and machinery involved and material consumed should be calculated. Such work is primarily calculated as explained in Chap. 12.

2. *Impact Cost*

Whenever a change is incorporated in the planned scope of work, it can affect the other unchanged work as well. Impact cost is the result of impact on other activity, delay costs, acceleration cost (overtime, etc.), and lost productivity or disruption costs.

To calculate the effect of changes on unchanged work, it is simply a matter of computing the additional hours of equipment and labor used or planned to be used along with additional materials that will be needed.

Most of the claims for loss and/or expense occur due to delay in completion of the works by any of the causes stated earlier. When the Contractor delays the project, the Employer can recover the loss by applying liquidated damages. However, when delay is not caused by the Contractor, then the Contractor is entitled to submit a claim, which usually is based not on delay in completion of the works but on the fact that the regular progress of those works has been disrupted.

It may be noted that the Contractor must prove that a loss has been suffered and that all possible steps had been taken by the Contractor to mitigate the effect of the delay, e.g., by way of reducing labor force, removal of machinery, and utilizing them productively elsewhere on site. The law does not allow a plaintiff to recover damages to compensate him for loss which would not have been suffered if he had taken reasonable measures to mitigate the loss.

When the work is prolonged or disrupted, the Contractor cannot achieve the productivity that was originally anticipated. It may be necessary for the Contractor to employ additional labor or work overtime, use extra materials or hire extra plant and equipment, simply in order to achieve the same result. With a prolonged duration claim, the cost of materials and labor may have escalated. In this way, many activities which are affected will be considered in pricing a claim.

Productivity or efficiency can be defined as

$$= \frac{Amount\ of\ work\ produced}{Amount\ of\ time\ (man-hours)} \text{ or } \frac{Production\ Output\ (say\ 200m\ of\ pipe)}{Resource\ Input\ (10\ Labor-hours)}$$

Loss of productivity or efficiency can be calculated using several methods. However, the "measured mile analysis" is widely acknowledged as the most acceptable method for calculating lost productivity costs. The analysis compares identical tasks in impacted and non-impacted periods of the project to estimate the productivity loss caused by the impact of a known series of events.

Generally, a productivity or efficiency claim seeks the increased labor cost. Typically, each area of lost productivity is determined by comparing productivity of the work performed under actual impacted conditions versus work rates achieved under normal working conditions. Once the area of lost productivity is determined, the damages are calculated for each individual item of work or task where productivity is lost. It is, therefore, very important to keep detailed time record when the project is disrupted. The record of similar performance on other projects will also help.

A simple example to understand the measured mile concept is provided here:

a. Measured mile: Assume the non-impacted or least impacted activity took 1000 man-hours to install 10 miles of PVC pipe.

   Hence, the productivity of period under normal conditions = 100 man-hours/mile

b. Impact: 640 man-hours were used to install 4 miles of PVC pipe.

   Hence, the productivity of the period that has been impacted = 160 man-hours/mile

c. Loss of productivity:

$$\begin{aligned} \text{Loss of Productivity} &= \text{Impacted conditions} - \text{Normal conditions} \\ &= 160 - 100 = 60\,\text{man-hours / mile} \\ &\text{or } 60 \times 4\,\text{miles} = 240\,\text{man-hours} \end{aligned}$$

It may be noted that the compensation because of impacts due to delay, inefficiency, disruption, or acceleration shall not include payment for profit. The intent of paying monetary damages is to pay the Contractor for actual compensable, quantifiable, and verifiable damages. This generally, does not include payment for profit. A Contractor is only entitled to earn a profit for constructing the project through producing quantities of work, which means actual work under contract, and changes in the work, but not for damages. OPSS provincial defines change in the work as the deletion, extension, increase, decrease, or alteration of lines, grades, quantities, etc. but does not consider damages as a change within the specified/contracted work.

3. *Overheads & Profit*

   In addition to actual and impact costs, the Contractor is also entitled to a reasonable allowance for site and head office overheads (indirect costs) and profit. Usually they are paid in the form of fixed percentage markups. These markups should be either specified in the Contract or agreed in writing by the parties after the contract has been awarded. It helps in avoiding the problem of computing actual additional overhead and costs for each change. The combined rate for overheads and profit employed by various organizations is 15%. However, sometimes separate percentage markups are also used depending on the nature of the claim.

Site overheads Site overheads usually include the expenditure on supervisory and administrative site staff and site accommodation including all water, electricity, telephone/fax charges, etc. It also includes construction equipment and tools along with necessary supplies for upkeep, running, and maintenance costs. If a fixed markup is not stated in the contract, site overheads can be computed from the Contractor's contemporary records.

Head office overheads Head office overheads usually include office rent, telephone/fax, electricity, documentation, office staff salaries, taxes, insurance if applicable, etc. Since these overhead expenses are an indirect expense which benefits the entire project, it must be allocated to all components on a shared basis.

As described above, it is always advisable to negotiate a fixed percentage markup, which should be added to actual and impact cost including overhead and profit. However, there are several methods that can be adopted to compute head office overheads.

Hudson formula This formula includes the head office overheads and profit combined:

$$\frac{h}{100} \times \frac{c}{cp} \times pd$$

where h = the head office overheads and profit including in the contract
c = the contract sum
cp = the contract period in weeks
pd = the period of delay in weeks

Eichleay formula For delay claims, this formula computes head office overheads on a daily rate basis. The Contractor is reimbursed that amount of overhead for each day of delay that occurred. This formula assumes that almost all overhead is fixed and evenly distributed across all projects. The formula is developed in three stages:

- *Step 1: Calculate the amount of overhead allocable to the contract*

$$\frac{\text{Contract billings}}{\text{Total billings for actual contract period}} \times$$
$$\text{Total overhead for contract period} = \text{Allocable overhead}$$

- *Step 2: Use the amount of overhead allocable to the contract to calculate a daily overhead amount*

$$\frac{\text{Allocable overhead}}{\text{Actual days of contract performance}} = \text{Daily overhead}$$

- *Step 3: Use the daily overhead amount to calculate unabsorbed overhead*

$$\text{Daily overhead} \times \text{number of days of delay} = \text{unabsorbed overhead}$$

4. *Other Costs*

Other costs usually include interest, lost anticipated profits, claim preparation costs, legal and professional fees, and others. Generally, these costs are non-compensable.

15.7 Claim Avoidance

Understanding the cause of claims as discussed above along with careful planning of how to deal with them will help to avoid or at least minimize the occurrence of claims.

1. *Contract Documents*: Always use standard forms of Contract Documents and specifications. One-sided or Owner-drafted Contract Documents are not advisable for both the parties, as they generate more claims instead of preventing them. However, it may be noted that usually standard forms of contract also need some modifications as many items are not provided in the standard forms of contract and need proper clarification to avoid confusion and claims. Make sure following clauses are provided in the contract:
2. *Contractor's Markup*: Ensure that Contractor's markup is clearly defined in the contract, for example:

   "Contractor's markup (overhead and profit combined) on work and services completed by its own forces shall be 15% and on work and services completed by its Subcontractor's shall be 5%," and similarly, a clause shall be added for Subcontractor's markup also.

- Ensure to clarify that the markup claimed by the General Contractor on his Subcontractor shall be on work done only, excluding markup; similarly markup claimed by a Subcontractor on his Subcontractor shall be claimed on work done only, excluding markup. In most cases, it is usually claimed on the total amount including markup. The limit on subcontractors of subcontractors should be clarified. Better to specify total figure of markup for a General Contractor and his subcontractors.
- The equipment rental rates provided by OPSS 127 [10] include direct and indirect costs and hence make sure no markup is added to the rental charges claimed.
- Sometimes contractors add additional bonding costs into their quotation, which as per Industry practice are not acceptable. Better to add a clause, "The Contractor shall (& subcontractor) not include any additional Bonding costs in their quotation for any Change Order work that may be required or performed."
- Often Contractors claim hours for superintendent on site in their claim for additional or extra work done on time and material bases. This is called "double-dipping" or requesting payment for the same cost twice, as supervision is part of indirect costs and is already covered in the agreed markup (overhead and profit) in the contract. However, to avoid this confusion and bad practice, it is better to add clause, such as:

  The mark-up provided in the contract shall constitute the only compensation the Contractor and/or Subcontractor shall be entitled to for any and all overhead, profit, incidental and administrative costs whatsoever related to the change, extra work or additional work, including but not limited to costs relating to superintendence and supervision, shop drawing production, estimating, site office, and home office expenses, workers tools, temporarily facilities, and controls, and coordination of any and all work related activities

3. *Standby Time*: OPSS Muni *100* [11] provides that "the Owner shall pay the Contractor for standby time of equipment at 35% of OPS 127 [10] rate or 35% of the invoice price whichever is appropriate". Some standard forms allow 50% of the invoice price for standby time of equipment. The Owner shall also pay for labor and/or operators who cannot be otherwise employed during the standby period.
4. Be familiar with the contract; read every article and all requirements of the contract thoroughly.
5. Carry out the contents of the contract faithfully and do what the contract requires.
6. Don't proceed with any variation, without proper documentation.
7. Use most appropriate type of contract delivery system.
8. Avoid incomplete, inaccurate, and poorly defined design details.
9. Define the scope of work in detail, so as to control what is and is not included in the project. Avoid ambiguities in the contract. When drawings and specifications have two reasonable interpretation, it is said to be ambiguous. Since Contract Documents are prepared by the Owner through the Consultant, any ambiguity in the interpretation of the contract will be interpreted in favor of the Contractor. Ambiguities should be resolved in a fair manner with the involvement of the Contractor to prevent delay claims.

The item description shall be provided clearly or in detail either in the specifications or in the form of tender/bill of quantities. In one of the GO transit station project, the description of an item provided in the form of tender was "Remove and dispose of canopies to Platform." Since the description of the item was not complete and was not further explained in the specifications what was incurred under this item, the Contractor claimed for removal of grade slab concrete of the canopy, which in fact was the part of overall canopy structure. It is therefore important to provide a clear description of what is required explaining clearly what is covered under the scope of that particular item.

Usually, a list of items (schedule of rates) is provided with the Form of Tender for Contractors to quote rates for any additional work if required. Make sure this list should address as many items as possible which would be helpful in pricing additional/extra items. An overly robust provisional rates section can invalidate the (perceived or actual) integrity of the contract. It serves to acknowledge that the Owner/Consultant missed a lot and anticipate a lot of changes. Also, if of large financial magnitude, it allows a Contractor to unbalance his bid and secure the provisional funds whether the items are used or not.

10. Prepare a realistic and more detailed schedule with related work breakdown structure (WBS) that fits the required dates in the contract and distribute it to all the related parties to the contract. so that everyone knows what is expected of them and can plan their work effectively and economically.
11. Allocate project risks fairly, sensibly, and equitably among the parties to the contract.
12. A major source of conflict is differing site conditions. A timely and thorough subsurface investigation, to know as much as possible about the site, by the Owner will minimize misunderstandings and disputes resulting from extra and additional works; however, if the investigations are inadequate, it will adversely affect the Owner's proposed budget. On a recent project, the dewatering volumes investigated by the geological consultants were underestimated which resulted in huge claims. The lesson learned was to allocate dewatering risk to the Contractor being the party that is best able to manage the risk and who has a duty to mitigate the damages, by adding the requirements of a PTTW (permission to take water) from the Ministry of Environment (MOE) in the scope of work. The condition can read as:

    The Contractor shall be responsible for obtaining the MOE permit for ground water lowering if deemed required prior to the start of dewatering works. All costs associated with obtaining the required permit including preparing and submitting required MOE documentation shall be paid for by the Contractor. If necessary, the Contractor shall use and pay for the services of a qualified geotechnical Consultant to prepare and submit the required documents on behalf of the Contractor and Owner.

13. With respect to differing site conditions mentioned above, contractors must review Owner provided geotechnical information carefully prior to submit bids. If the information is included as a "guide," then the contractor might be at

risk if he relies completely on this information based on a court's ruling with respect to a claim of *Stuyvesant Dredging Company v. United States* under Army Corps of Engineers contract for maintenance dredging of the Corpus Christi Entrance Channel in Texas [12]. Under this case, the Contractor claimed that during the performance of the work, they encountered increased density of materials which impacted their productivity resulting in an increase in cost. The court examined the case and denied the claim; its ruling was "the six average density readings were identified to be guide only – did not reach the level of estimates and were clearly not facts upon which plaintiff could rely [13]."

14. Better site supervision and control by qualified, experienced, and trained personnel will help to resolve issues at the initial stage minimizing the occurrence of claims and conflicts.

Reference and Further Reading

1. CCDC-2. (2008). Canadian Construction Document Committee – Stipulated Price Contract.
2. Texas Department of Transportation (Third Quarter 2007) – The disputes and claims analysis process – Construction and material tips. Published Quarterly by the Construction and Bridge Division.
3. Morrison-Knudsen Co. v. British Columbia Hydro & Power Authority (No.2) (1978) 85 D.L.R (3d) 186 (BC. Ct App).
4. W.A. Stephenson Construction (Western) Ltd. v. Metro Canada Ltd. 27(1987), 27 C.L.R. 113 (B.C. S.C.).
5. Levin, P. (1998). *Construction contract claims, changes & dispute resolution* (2nd ed.). VA, USA: American Society of Civil Engineers Press.
6. DLA Piper Law Firm. (2013). Asia Pacific projects update- An Article on "Concurrent Delay".
7. FIDIC. (1999). Conditions of contract for construction. Red Book: Author FIDIC (International Federation of Consulting Engineers).
8. Federal Acquisition Regulation, §52.236–2, *Differing Site Conditions* (APR 1984). (48 FR 42478, Sept. 19, 1983, as amended at 60 FR 34761, July 3, 1995).
9. Fullerton & Knowles; Attorneys and Counsellors at Law in Virginia, Maryland, Pennsylvania and the District of Columbia- Construction Law Survival Manual- Ch 5 – Changes, delays and other claims.
10. OPSS 127. (2015). Ontario provincial standard specification – Schedule of rental rates for construction equipment, including model and specification reference.
11. OPS. (2006). General conditions of contract – Ontario Provincial Standards for Roads & Public Works- OPSS. MUNI 100.
12. Stuyvesant Dredging Company, Plaintiff-appellant, v. the United States, Defendant-appellee, 834 F.2d 1576 (Fed. Cir. 1987).
13. Collins, S. A., & Zack, J. G, Jr. A technical article on "Changing Trend in Risk Allocation – Differing Site Conditions".

Chapter 16
Suspension and Termination of Construction

16.1 Introduction

Suspension occurs on a construction project when an Owner instructs a Contractor to temporarily stop work on all or a part of the project. Termination occurs when an Owner instructs a Contractor to permanently stop the performance of work and leave the site. Standard contracts mostly specify each party's rights, obligations, and remedies for suspension and termination in certain circumstances. These clauses most commonly deal with breaches of specified contractual obligations. There may be rights to terminate the contract in other situations too, such as the occurrence of a force majeure event or failure to pass a tests.

Suspension and termination on construction projects often result in claims and disputes; therefore, the decision to proceed with either option should not be taken lightly. The resulting consequences, both practical and legal, should be considered before notice is given to the Contractor. Suspension or termination steps should not be taken without considerable discussion and attempt to overcome any contractual problems and rectify the situation. There will be significant financial consequences for the Owner if the Contractor objects to termination and the dispute adjudication board (DAB) or arbitrator later decides that the Owner was not entitled to terminate [7].

16.2 Contract Termination and Discharge

While a contract is wholly or partly unperformed, it may be terminated due to certain reasons. It is desirable for the contract to include a termination clause in order to provide for an orderly and equitable procedure in the event of circumstances, which make it prudent or necessary to terminate the contract. Before the contract is terminated, it would be in the interest of both parties to restore to other measures or remedies provided within the contract.

A. Surahyo, *Understanding Construction Contracts*,
DOI 10.1007/978-3-319-66685-3_16

A contract is said to be discharged when parties "discharge" or perform their rights and obligations as required under the contract. Performance thus leads to the end of the contract. Nonperformance of the required rights and obligations by either or both parties leads to the termination of a contract. Contracts may be terminated or brought to an end (discharged) in a number of ways, such as:

1. *By performance*: Full and satisfactory performance by both parties to their obligations under the contract within the time and in the manner prescribed. Thus the parties are discharged and the contract comes to an end.
2. *By breach of contract/default*: When a party fails to perform an obligation under the terms of a contract, specifications, or delivery schedule, the defaulting party is said to have breached the contract. However, not every failure to perform an obligation amounts to a true breach, as there are a number of excuses for nonperformance. Nonperformance, faulty performance, slow progress, and non-compliance of instructions are examples of material default by the contractor for which the Owner has right to take action against the Contractor under the contract.
3. *Fundamental breach*: A breach that goes to the root of the contract means when the party in breach has committed a serious (or fundamental) breach of the contract and deprives the innocent party of the substantial benefit, he is entitled to under the contract.

For example, Contractor's repeated non-compliance to the Consultant's instructions to rectify the defect is a fundamental breach of the contract. In a case *R.F.M. Electric Ltd. v. the University of British Columbia and Martina Enterprises*, it was held that [4]:

> R.F.M. had failed repeatedly to comply with the specifications and to comply with the electrical General Contractor/Consultant's directions. The court concluded that such non-compliance constituted a fundamental breach of the contract.

4. *Repudiation breach*: When one party to a contract verbally or by conduct tells the other party that he has no intention of performing his obligations, he has repudiated the contract. Without justification, repudiation constitutes a fundamental breach. The example includes abandonment of the site or removal of plant by the Contractor, employing other contractors to carry out the same work.
5. *By mutual agreement*: Commonly used in the commercial contracting environment. Terminating by mutual consent is a bilateral agreement indicating that the parties no longer wish to be bound by the contract and terminating both parties' respective rights and obligations [3].

This usually happens where the Owner has requested full or partial termination of a contract; the Contractor has incurred minor or no expenses and is willing to forego a claim, and the matter may be settled at no cost to any party [6].

Mutual contract termination can also happen where the contract is no longer being followed or if the parties have ceased business operations, or if the contract can no longer be faithfully performed, the parties may wish to formally terminate the agreement in writing.

6. *By unforeseen circumstances (force majeure):* Unforeseen circumstances beyond the control of either party make it impossible to perform duties or obligations stated in the contract. If this happens, the contract is termed as "frustrated" and is discharged by such frustration.

Frustration occurs whenever a contract, after its formation, becomes impossible to perform without default of either party; the doctrine is often called subsequent or supervening impossibility, and its effect is that the parties are released from their contractual obligations [5]. A common example is that, during a renovation project, the house burns down without fault of the parties.

7. *By operation of law:* A contract may be discharged independently of the wishes of the parties, i.e., by operation of law. This includes discharge:
   - By death
   - By insolvency
   - By unauthorized amendment of the terms of a written contract
   - By rights and liabilities becoming vested in the same person

The most common manner by which a contract is brought to an end is the performance by the parties of their respective obligations under that contract. The promises are performed and the contractual obligations are satisfied. However, the most uncommon circumstances of leading the contract to an end are the result of breach of contract. When a contract has been broken without sufficient excuse or justification, the party who suffers by such breach (innocent party) is entitled to receive from the defaulting party a compensation for any loss or damage caused by such breach.

The most common remedy for breach of contract is financial damages for the loss incurred; however, for serious breach, the remedy of termination is available in contracts. The damages awarded in an attempt to re-compensate the actual loss sustained by reason of the breach of contract may include liquidated damages for delay in completion of works, extra cost for repair of defective works, etc.

16.3 Termination by the Owner Under FIDIC Conditions of Contract [1]

FIDIC clause 15 describes the circumstances that may lead to a termination of the Contract by the Employer either as a result of a default by the Contractor or for convenience. It further describes the procedures that need to be followed and the financial arrangements that will apply.

However, under FIDIC [1] there are also some additional Sub-clauses according to which the Employer is entitled to terminate the contract in certain circumstances:

- Sub-clause 9.4(b): failure to pass tests on completion
- Sub-clause 11.4(c): failure to remedy defects
- Sub-clause 19.6: optional termination payment and release (force majeure)
- Sub-clause 19.7: release from performance under the law

At initial stage of a potential termination stage, FIDIC [1] Sub-clause 15.1 specifies that if the Contractor fails to carry out any obligation under the contract, then the Engineer should issue a notice to the Contractor for rectification of the failure situation within a certain reasonable time period. FIDIC Sub-clause 15.2 further summarizes the events which under the conditions constitute events of default by the Contractor:

(a) When the Contractor fails to comply with Sub-clause 4.2 (performance security) or with a notice under Sub-clause15.1 (notice to correct) as mentioned above. Per Sub-clause 4.2, the Contractor will be considered defaulted. Default can occur if the Contractor:

   - Fails to provide the performance security within the required time
   - Fails to ensure it is valid and enforceable for the period required
   - Fails to extend its validity as required

(b) When the Contractor has repudiated or abandoned the contract
(c) When the Contractor without reasonable excuse fails to commence the work in a reasonable time in accordance with Sub-clause 8; fails to make good the rejected defective plant, materials, or workmanship by the Engineer under Sub-clause 7.5; or fails to comply with the instructions of the Engineer regarding rectification of defective works or materials within 28 days after receiving notice
(d) When the Contractor subcontracts all the works or assigns the contract without the required agreement under the contract
(e) When the contractor becomes bankrupt or insolvent
(f) When the Contractor or his Personnel, agents, or subcontractors offers any bribes, gifts, gratuities, or commission as an inducement or reward to a person for doing or allowing any action in relation to the contract or to gain favor or disfavor to any person in relation to the contract except any lawful inducements and rewards to Contractor's personnel

In the event of specified defaults by the Contractor, the Employer may terminate the Contractor's employment under the contract for Contractor's default by giving 14-day notice as set out in the above mentioned clause except in the case of subparagraph (e) or (f) where the Employer can terminate the contract immediately by issuing such notice.

16.3.1 Costs and Risks of Terminating the Contract

Once the contract is terminated, the innocent party is entitled to damages for losses incurred based on the conditions of the contract or to a quantum meruit remedy for reasonable value of work performed.

Contractor's default If the Owner terminates the Contractor's employment under the contract due to Contractor's default, the measure of the Owner's damages is usually the cost of completing the scope of the works and rectifying any defects, after taking into account the unpaid balance of the contract price. The Owner may complete the works himself or may employ any other Contractor to complete the works. For this purpose, all the following will apply according to FIDIC [1] Conditions of Contract:

- As per FIDIC Sub-clause 15.2, the Contractor must leave the site as soon as reasonably practicable within the provided notice period and deliver any required equipment, temporary works, materials, and documents to the Engineer so as to enable the Owner to complete the works. This subparagraph also contemplates the termination notice including instructions concerning safety and the assignment of subcontracts.
- The Engineer, on behalf of the Employer, as per Sub-clause 15.3, should determine the amount of work done by the Contractor at the time of termination and the value of any goods and Contractor's documents as well as any other amount due to the Contractor for work done in accordance with the contract.
- As per FIDIC Sub-clause 15.4, the Employer is not liable to pay the Contractor any further amounts in respect of the contract until the cost of completing the works; remedying of any defects, losses, or damages for delay in completion (if any); and other expenses incurred by the Employer as a result of the Contractor's default have been established.
- The Contractor shall be entitled to receive only such sums (if any) which are due to him duly certified by the Engineer after deducting the above mentioned amounts. If such amount would exceed the sum which would have been payable to the Contractor on due completion by him, the Contractor shall be liable to pay the Employer such excess amount, and it shall be deemed a debt due by the Contractor to the Employer.
- The Employer shall then release the Contractor's equipment and temporary works. The Contractor shall promptly arrange their removal, at the risk and cost of the Contractor. However, if the Contractor would fail to make a payment due to the Employer, these items may be sold by the Employer so as to recover this payment. Any balance shall then be paid to the Contractor.
- The Contractor must assign to the Employer his rights and benefits in all contracts concerning the works, warranties and undertakings, bank guarantees, and retention held by the Contractor, with effect from the date of termination of his employment under the contract.

16.3.2 Termination by the Employer for Convenience (Without Cause)

As per FIDIC [1] Sub-clause 15.5, the Employer is also entitled to terminate the contract at any time for convenience, that is, without any default on the part of the Contractor. The termination shall take effect 28 days after the Contractor receives such notice or the Employer returns the performance security. Under this clause, the Employer shall not terminate the contract to execute either himself or through another Contractor.

After termination under Sub-clause 15.5, FIDIC [1] provides that the Contractor shall proceed in accordance with Sub-clause 16.3 and shall be paid in accordance with Sub-clause 19.6. Sub-clause 16.3 requires that the Contractor shall promptly cease all further work except any work instructed by the Engineer with respect to safety of works; hand over all plant, material, documents, and other works for which the contractor has already received the payment; and remove all his other equipment except as necessary for safety requirements and leave the site.

Under Sub-clause 19.6, the Engineer shall determine the value of work done and issue a payment certificate. The value of work should include:

(a) Payment of the work done prior to termination
(b) Cost for plant and materials ordered for the work
(c) Other costs incurred by the Contractor in the expectation of completing the works
(d) The cost of removal of temporary works and Contractor's equipment from the site and the return of these items to the Contractor's works in his country
(e) The cost of repatriation of the Contractor's staff and labor

Sub-clause 19.6 also provides that if the progress of work is prevented for a continuous period of 84 days or for multiple periods of more than 140 days due to force majeure, then either party may give notice of contract termination.

16.4 Termination by the Owner Under CCDC-2 Conditions of Contract [2]

CCDC-2 Sub-clause 7.1 also authorizes the Owner to terminate the contract by giving such notice under certain conditions. These conditions are summarized as under:

- If the Contractor becomes bankrupt
- If the Contractor neglects to execute the work properly or fails to comply with the contract requirements substantially

In the second condition, following the Consultant's justification in writing that sufficient cause exists to justify the termination action, the Owner shall give notice in writing to the Contractor to correct the default in 5 working days. If the default cannot be corrected in the 5 working days or in such other agreed time, the Contractor still shall be considered in compliance of the Owner's instructions if he starts correction of the default and submits an acceptable schedule for correction (and corrects the default in compliance with the contract terms and that schedule).

If the Contractor fails to correct the default in the specified agreed time period, then the Owner may correct such default and deduct the cost from the Contractor's due payments in agreement with the Consultant or terminate the contractor's right to continue with the overall work or part or terminate the contract.

Once the Owner terminates the contract, then per Sub-clause 7.1.5 of CCDC-2, the Owner shall be entitled to:

- Take over the work and products including equipment of the Contractor subject to the rights of third parties. Finish the balance of work in a reasonable time by any suitable method.
- Stop any further payment to the Contractor until a final certificate for payment is issued.
- Charge the Contractor for completing the work as certified by the Consultant including the Consultant's service charges or any corrections required under warranty.

CCDC-2 [2] Sub-clause 7.1.6 provides that the Contractor's obligation with respect to quality, correction, and warranty of the work done by the Contractor up to the time of termination shall continue after the contract termination.

CCDC-2 [2] has no such clause according to which the Owner can terminate the contract at his convenience or without cause like FIDIC [1].

16.5 Termination by the Contractor Under FIDIC Conditions of Contract [1]

Under Sub-clause 16.2 of FIDIC, the Contractor is also entitled to terminate the contract based on following reasons:

(a) If the Employer fails to provide evidence of financial arrangements as per Sub-clause 2.4 within 42 days after giving notice by the Contractor under Sub-clause 16.1
(b) The Engineer fails to issue a payment certificate within 56 days after receiving the statement and supporting documents by the contractor.
(c) The Contractor does not receive the interim payment for the work done within 42 days after the required expiry date.
(d) The Employer fails to perform his obligations substantially as required under the contract.
(e) The Employer fails to perform agreement clause 1.6 or assignment clause 1.7.
(f) If the suspension affects the entirety of the works for long duration as per Sub-clause 8.11.

(g) The Employer becomes bankrupt.

The Contractor must give 14-day notice before he terminates the contract; however in the case subparagraph (f) and (g), the Contractor can terminate the contract with immediate notice.

Employer's default If the Contractor terminates the contract due to default of the Employer, the Contractor is entitled to his damages under the contract. The following will apply to the Contractor's termination under FIDIC [1]:

- As per Sub-clause 16.3, the Contractor must stop further work, leave the site as soon as reasonably practicable, and remove all plant, equipment, and amenities he had brought onto the site for constructing the works except as necessary for safety. Hand over to the Engineer all those items for which the Contractor has received payment.
- Under Sub-clause 16.4, the Employer is obliged to make payment to the Contractor for all works executed fully or partially prior to the date of termination including materials supplied or ordered for the works as defined in Sub-clause 19.6. Pay the amount of any loss or damage to the Contractor arising out of or in connection with or by consequence of such termination, and return the performance security to the Contractor.

16.6 Suspension of Construction Under FIDIC Condition of Contract [1]

Under most standard forms of contract, express provisions are made that allow an Employer to order the temporary suspension of the work. This may be exercisable on specified grounds or more widely at the Owner's convenience and without any duty to give reasons. However, while contractual power to suspend is often wide, the use in practice is often limited because of the likely delay to completion and high costs that the Employer may have to pay as a result.

Sub-clause 8.8 of FIDIC has a provision whereby the Engineer is authorized to suspend the work or any part, thereof, for a certain period and may notify the cause for suspension. During the suspension period, the contractor is responsible to protect the works from any loss or damage.

If the Contractor is not responsible for the suspension and suffers delay or incurs cost due to the Engineer's instructions under Sub-clause 8.8, he should then provide such notice to the Engineer as per Sub-clause 8.9. The Engineer, after consultation with the Employer and the Contractor as required under Sub-clause 3.5, should determine the extension of time required and/or extra costs the Contractor shall be entitled to receive. However, if the progress of work is stopped due to the Contractor's actions/faults, then he will not be liable for any extension of time or extra costs.

As such suspension may result in serious losses for a Contractor, a provision is usually made in the standard forms of contract that allows a Contractor to terminate the contract if suspension is prolonged and continues beyond a certain specified time period. FIDIC Sub-clause 8.11 includes such a provision and states that if the progress of work or any part thereof is suspended for more than 84 days then, unless the Contractor is responsible for the suspension, the Contractor may give notice to the Engineer requiring permission to proceed within 28 days. If permission is not granted within the notified period, the Contractor may treat the suspended part of the work as omitted, or where all the work has been suspended, treat the contract as repudiated and terminate his employment under the contract in accordance with the provision of Sub-clause 16.2.

However, under FIDIC Sub-clause 16.1, the Contractor is also entitled to suspend the work as mentioned above under subparagraph (a), (b), and (c) of "Termination by the Contractor." The Contractor is entitled to obtain a time extension for any delay and compensation for his extra costs. The Contractor should resume normal work as soon as is reasonably possible (before the notice for termination is due), when the cause of suspension has been removed.

Because of the legal consequences that follow from suspension, it is important for contractors to ensure before work is suspended that the Engineer has provided written instructions/notice. If a verbal instruction is given, a written instruction should be obtained. Failure to obtain the required notice may create difficulties in claiming extension of time and/or costs.

16.7 Termination and Suspension by the Contractor Under CCDC-2 [2]

Like FIDIC [1], CCDC-2 [2] also has provisions whereby the Contractor is authorized to terminate or suspend the contracted work. CCDC-2 Sub-clause 7.2 provides that the Contractor can either terminate or suspend the work if:

- The Owner becomes bankrupt.
- The work is suspended or delayed for a period of about 20 working days or more by court or other public authority order at no fault of the Contractor or his employees or any one engaged by the Contractor.
- The Owner is in default of such contractual obligations as providing evidence of financial arrangements required under the contract, or the Consultant fails to issue a progress payment certificate as per Sub-clause 5.3, or the Owner fails to pay amounts due to the Contractor as certified by the Consultant or awarded by arbitration/court, or the Owner violates the contractual obligations substantially and confirmed by the Consultant in writing to the Contractor except for Owner's financial information as per Sub-clause 5.1.

In the last case, the Contractor must give 5 working days' notice in writing to the Owner prior to taking any action, in order to allow the Owner an opportunity to remedy the default. Following termination of the contract, the Contractor shall be entitled to payment of all work performed including with profit and any material supplied or ordered. The Owner has to pay for resulting loss or damage sustained due to termination of the contract.

The procedure set down in the Contract Documents for termination should always be followed strictly. If a procedural step required by the termination clause is not followed, the party seeking to end the contract runs the risk of being in default by not following the contract. As a general rule, the termination process must be clear and must be performed in such a way that evidence of actions leading to the termination is readily available to the defaulting party.

In the event of breach of contract, a party that suffers a loss must take reasonable steps to mitigate or reduce the amount of damages suffered. For example, if the party does not take care in mitigating damages, his conduct will be taken into account when the court is fixing the damage award.

References and Further Reading

1. FIDIC. (1999). *Conditions of contract for construction.* Red book: Author FIDIC (International Federation of Consulting Engineers).
2. CCDC-2. (2008). *Canadian construction document committee – Stipulated price contract.*
3. Wilkinson, J. W. 2010. Certified professional contracts manager study guide. Published by National Contract Management Association (NCMA) USA.
4. R.F.M. Electric Ltd. v. University of British Columbia- Between R.F.M. Electric Ltd., Plaintiff, and The University of British Columbia and Martina Enterprises, Defendants. [1992] B.C.J. No. 1810- British Columbia Supreme Court.
5. Insite Law magazine (Nov 28, 2013), Daily Online Law News and Law Blogs.
6. Public Works and Government Services Canada; 8.135.20. Termination by Mutual Consent- Buyandsell.gc.ca.
7. Rebecca Saunders- FIDIC: Termination by the Employer under the Yellow & Red Books. Fenwick Elliot: The Construction & Energy Law Specialists.

Chapter 17
Construction Disputes

17.1 Introduction

Unexpected situations, leading to disputes, are a normal feature of every construction project because conditions encountered in practice are usually different from those planned or foreseen. Construction disputes arise due to a series of factors that combine in various ways to produce arguments, disagreements, and finally disputes. These disputes may be of varied nature like technical, financial or quality related, insurance related, duration related, or staff/labor related, etc.

Most disputes appear during construction and several techniques are available to help contracting parties to come to some form of settlement. A primary goal in any claim or dispute is to maintain control. One aspect, which is often controllable, is the choice of a claim or dispute settlement method. Knowledge about settlement options and their respective advantages helps to maintain that control. Typically, claims and disputes can be settled by one or more methods; however, we will address CCDC-2 [1] and FIDIC [2] conditions of contract approaches in this regard.

17.2 CCDC-2 Dispute Resolution Procedure

Dispute resolution between the parties to the contract is addressed under Clause-8 of CCDC-2 [1]. The suggested structure for dispute resolution is negotiation, mediation, and arbitration. In general, being the Contract Administrator of the project, the Consultant has traditionally been the primary Arbiter with respect to disagreements or disputes between the Owner and the Contractor. At the initial stage, as required by CCDC-2 Clause 2.2; the Consultant will interpret and investigate the dispute in an unbiased, fair, and professional manner, consistent with the intent of the Contract Documents and convey his decision to both parties in writing.

A. Surahyo, *Understanding Construction Contracts*,
DOI 10.1007/978-3-319-66685-3_17

17.2.1 Negotiation

If any one of the parties is unsatisfied with any particular finding made by the project Consultant, the unsatisfied party, within 15 working days of the receipt of the Consultant's findings, must send a notice in writing of the dispute to the opposing party and the Consultant. That notice must contain the particulars of the specific matter in dispute, as well as the relevant provisions of the contract. The sending of a notice serves to invoke the next formal step in the dispute resolution process, "negotiation." Within 10 working days of receiving the notice in writing of the dispute, the opposing party must send a notice in writing of reply to the dispute. The parties must then sit down and make all reasonable efforts to resolve their dispute by amicable negotiations before moving on to the next step of mediation in the process.

17.2.2 Mediation

If the dispute is still unresolved after the 10-working day process of negotiation has been exhausted, the parties must ask the Project Mediator to step in to assist them in resolving their dispute. CCDC-2 [1] provides that the mediated negotiation shall be conducted in accordance with the "Rules for Mediation of Construction Disputes" as provided in CCDC-40 [3]. It further suggests that the parties shall appoint a Project Mediator either within 20 working days of the date of the award of the contract or within 10 working days after either party makes a written request by notice in writing that a Project Mediator be appointed.

The Project Mediator will conduct mediation for resolution of the dispute, meeting with both sides individually prior to joint discussion on a "without prejudice" basis. The Mediator's decision is not binding on either party, because the Mediator's role is to motivate the parties to communicate with each other while keeping all disclosures, as well as any resultant settlement, confidential. If the dispute is not resolved within 10 days or within such a period agreed to by the parties, the Mediator will terminate the mediation by issuing notice to the Owner, the Contractor, and the Consultant.

17.2.3 Arbitration

Once the mediated negotiations are terminated, either party may refer the dispute within 10 working days to be finally resolved by the arbitration under the Rules for Arbitration of Construction Disputes as provided in CCDC 40 [3]. CCDC-2 [1] provides that the arbitration should be conducted in the jurisdiction of the "Place of the Work." After the expiry of 10 working days, the arbitration is not binding on the parties, and they may refer the dispute to the courts.

The procedures negotiations, mediation, and arbitration are further described under the following pages.

17.3 FIDIC Dispute Resolution Procedure

FIDIC [2] approaches the role of the Engineer in a different manner from that anticipated by the Canadian standard forms. While the Canadian standards include content addressing the role of the Consultant as an impartial administrator and arbiter in the first instance, FIDIC, the international suite of standard form contracts, expressly denies the impartial character of the role of the Engineer who is engaged by the Employer [4]. Accordingly, FIDIC, under clause 20.2, provides that disputes shall be adjudicated by an independent dispute resolution process employing a dispute adjudication board (DAB) in accordance with clause 20.4.

17.3.1 The Dispute Adjudication Board (DAB)

The DAB procedure is treated as a method of primary dispute resolution under FIDIC [2] conditions of contract. Detailed provisions and procedures for the DAB are included in Sub-clauses 20.2–20.4, the Appendix A1 "General Conditions of Dispute Adjudication Agreement," and the Appendix A2 "Procedural Rules."

The DAB is an independent panel of one or three suitably qualified people who are appointed at the beginning of the project so as to give decisions on any dispute. For the DAB to operate successfully, it is essential that the members:

(a) Should visit the site regularly so as to become familiar with the details of the project.
(b) Should keep themselves up to date with activities, progress, and problems at the site.
(c) Should be permitted access to all documentation and be allowed to attend such meetings as may be necessary to become and remain informed.
(d) Should encourage the parties to resolve the disputes mutually at the initial stage.
(e) Ensure decisions by DAB are implemented as early as possible.

As per general conditions for agreement under Appendix A1, a member is not permitted to be appointed as an arbitrator or to give evidence in any arbitration under the contract except it is agreed by the Employer, Contractor, and other members.

17.3.1.1 Appointment of DAB

As per the Sub-clause 20.2, the DAB is to be appointed within 28 days of the commencement date as stated in the appendix to tender. If the DAB comprises three members, each party shall nominate one member, for the approval of other party.

The chairman is then appointed or agreed by the parties with the consultation of selected two members or by the appointing authority named in the appendix to tender. Each party must ensure to nominate a truly independent expert with the ability and freedom to act impartially.

If the parties fail to agree on the nomination of DAB members, then the FIDIC [2] appendix to tender designates the president of FIDIC [2] as the appointing authority, who will nominate individual DAB members if requested to do so. The appendix to tender can be amended to designate a different appointing authority.

FIDIC [2] specifies that the DAB's decision must be implemented and shall be binding on both parties, but if either party is not satisfied with DAB decision, the original dispute can be referred to arbitration.

The members selected to a DAB must be suitably qualified, impartial, and accepted and trusted by both parties. They should have appropriate construction experience, including experience within claims and dispute resolutions and must be free from any prior relationships that could be seen to lead to bias or a conflict of interest. They should also carry the knowledge of contract interpretation with an understanding of the rights, obligations, and liabilities of the parties and knowledge of the DAB procedures.

Once the appointment of the members is accepted, each member of the DAB signs an agreement with both parties. Standard forms for the Dispute Adjudication Agreement for one person and three person DAB are included with the FIDIC conditions of contract [2]. The agreement refers to the "General Conditions of Dispute Adjudication Agreement" and the "Procedural Rules" which are attached to clause 20.

17.3.1.2 Termination or Replacement

FIDIC [2] Sub-clause 20.2 states that neither the Employer nor the Contractor can terminate the appointment of any member. The Sub-clause 20.2 further states that only by mutual agreement of both parties the appointment of any DAB member may be terminated. Additionally, in case any DAB member declines to act or is unable to act as a result of death, disability, resignation, or termination of appointment, then a replacement can be made with the mutual agreement of both the parties by appointing a suitable qualified person or persons.

The Sub-clause 20.2 also provides that the appointment of any member can be terminated by joint agreement of both the parties when the Contractor's discharge under Sub-clause 14.12 becomes effective that means when the Contractor's final accounts are settled and he receives back his performance security.

The General Conditions of the Dispute Adjudication Agreement under Appendix A1 provide that for termination of the Dispute Adjudication Agreement, the Employer and the Contractor jointly need to give 42-day notice to the member. Conversely, a member can also terminate the Dispute Adjudication Agreement, by providing such notice to the parties, if the Employer or Contractor fails to comply with any condition of the Adjudication Agreement.

17.3.1.3 Obtaining DAB Decision

FIDIC [2] Sub-clause 20.4 states that if a dispute of any kind whatsoever arises between the parties in connection with or arising out of the contract, it should be referred immediately in writing to the DAB with copies to the other party and the Engineer.

Having received the referral notice, the DAB is required to follow the procedural rules as stated in FIDIC Conditions. Both parties are obliged per the contract to make available to the DAB such items as further information, access to the site, and any appropriate facilities as the DAB may require for the process of making a decision. The DAB may conduct a hearing during dispute investigations from the Employer and the Contractor.

Sub-clause 20.4 further provides that the DAB shall give its decision within 84 days or other agreed period between all parties. The decision will be in writing and include an explanation of the reasons. The decision is required to be binding upon the parties unless it is revised by an amicable settlement or an arbitration award.

If either party is dissatisfied with the decision of the DAB, then it must issue a formal notice of dissatisfaction. The notice must be issued within 28 days of the DAB's decision and must state the reasons for the dissatisfaction. Any failure of the DAB to reach a decision by the due date will also be a valid reason for issuing a notice of dissatisfaction.

If no notice of dissatisfaction is served by either party within the 28-day period, then the decision will become final and binding on both parties. However, as per Sub-clause 20.7, if a party fails to comply with DAB decision, then the other party may refer the failure itself to arbitration.

17.3.1.4 Alternative to Sub-clause 20.4

The FIDIC guide for the preparation of particular conditions includes an alternative paragraph for Sub-clause 20.4, which enables the Engineer to be appointed as a member of the DAB. In this case the Engineer is required to follow the DAB procedures, acting fairly and impartially, and his fees will be paid by the Employer.

It is difficult to imagine that the Engineer can perform the DAB duties while still acting in his role on behalf of the Employer [5].

17.3.1.5 Payment to DAB Members

As per the General Conditions of the Dispute Adjudication Agreement, members shall be paid in the currency named in the agreement. The DAB member shall be paid on the basis of a fixed monthly retainer fee and a daily fee. The retainer fee will compensate each member for being available, for time spent reviewing, and for maintaining project documents, communications, and clerical work including all office and overhead expenses.

The daily fee covers payment for site visits including travel time of the member, the time involved in preparing decisions and reading submissions in preparation for a hearing, etc.

The payment made to the member should also cover any tax if applicable. In accordance with clause 20.2, the Contractor and the Employer each will pay one-half the payments made to the DAB member.

Payment for all fees and expenses, as specified in the Dispute Adjudication Agreement, are made by the submission of invoices by the member to the Contractor. The Contractor is obliged to pay all the member's invoices in full within 56 calendar days. The Contractor is reimbursed by the Employer by one-half of the invoiced amount. In the event of any failure of the Contractor to pay the amount required, then the Employer is required to pay the due amount. The Employer is thereafter reimbursed by the Contractor by one-half of the invoiced amount plus any collection costs and financing charges.

17.3.2 Amicable Settlement

Following issue of a notice of dissatisfaction under Sub-clause 20.4 as discussed above, FIDIC [2] Sub-clause 20.5 provides a 56-day period before either party can commence arbitration. This clause further recommends that the parties shall attempt to settle the dispute amicably prior to proceeding for arbitration but do not give guidance for any procedure to be followed.

Amicable settlement primarily involves three processes, such as negotiation, conciliation, and mediation and excludes arbitration or litigation, where a binding decision is imposed on the parties.

Under the amicable settlement, parties are brought to the negotiating table to discuss their problems, establish facts, clarify issues, and work out some settlement options. The goal is to reach an agreement based on shared understanding and joint efforts by the parties which is not imposed on them by the third party as is in arbitration.

Amicable settlement is more suitable for disputes within the construction industry, where all parties have worked hard to complete the project on time and within budget and usually hope to work together again in future.

17.3.2.1 Negotiations

Negotiation is the primary method of amicable settlement process and is probably the most flexible form of dispute resolution as it involves only those parties with an interest in the matter and their representatives, if any. Negotiation is the least costly and informal method of dispute resolution, allowing a high degree of control over issues and time factors.

Since negotiation is a voluntary process, the first and fundamental step to be taken is to confirm whether or not the other party or parties are interested in negotiations. In making such an assessment, it is important to take into account the following factors [6]:

1. The desire to resolve the dispute
2. Whether a negotiated solution is in the interests of any or all of the parties in question
3. The credibility of the other party
4. The willingness of the parties to establish or preserve a relationship
5. Whether or not there is a disparity between the parties to the extent that it would be impossible to bargain equally, i.e., there is a marked contrast between the parties in terms of the level of education or the resources of the parties
6. The desirability of using another form of alternative dispute resolution, such as mediation or arbitration
7. Proper authority to enter into negotiations and to reach an agreement or settlement

A successful negotiation requires each party to have a clear understanding of its negotiating mandate. If uncertainty exists regarding the limits of a party's negotiating authority, the party will not be able to participate effectively in the bargaining process. The dispute resolution should not be based on either winning or compromise criteria but should concentrate on satisfying as many as possible of the important needs and interests of the parties. During the negotiation, both parties need to remain cordial, tolerant, and patient. Each party has to leave some bargaining room and expect some give and take. The negotiations are considered successful when one party's best outcome is good for the other party. Former US Secretary of State, Henry Kissinger, said that each party to a negotiation has certain requirements and goals – and the art of negotiation is finding out what the other party's goals are and satisfy them.

The negotiations may be conducted between senior personnel of the parties; however, the presence of the Consultant is most important. The traditional role of the Consultant, as an impartial professional, includes interpreting and making initial decisions on disputes concerning the contractor and the Owner.

The most important item for a successful negotiator is adequate preparation and familiarity with the project. The negotiator must establish objectives before going to the meeting such as which item is to be compromised and to what extent and which items are not to be compromised or totally dropped. While starting discussions, it is always useful to start with the items which should be resolved easily, taking the heat off. Negotiations should be fair and honest. Detailed records of the negotiations must be kept.

17.3.2.2 Mediation

Mediation is basically a nonbinding dispute resolution process, where a neutral professional is requested to help the parties to reach a negotiated and acceptable settlement.

If the dispute cannot be resolved through negotiations, then the mediation procedures are implemented if the contract requires. For mediation, the parties select a sole neutral person/professional who encourages the parties to communicate, understand, and evaluate each other's viewpoint and agree to a settlement. The mediator basically helps the parties to discuss in detail the strengths and weaknesses of their cases and assists them in reaching their own mutually acceptable settlement of issues in dispute. The mediator does not decide for the parties but helps them make their own decision.

The South African Association of Consulting Engineers propose that the general philosophy of mediation should be that the disputes which arise in the execution of engineering works by contract can be settled by Engineer Mediators for other Engineers to avoid "mortal combat" resulting from arbitration or litigation [7]. Mediation is particularly useful when the disputing parties need or desire to maintain an ongoing relationship. Mediation allows parties to avoid the adversarial elements of litigation which often make it impossible to continue a productive relationship after the settlement.

The role of Mediator can be summarized as to [6]:

1. Encourage exchanges of information.
2. Help the parties understand each other's views.
3. Let the parties know that their concerns are understood.
4. Promote a productive level of emotional expression.
5. Lay out the differences in perceptions and interests.
6. Identify and narrow issues.
7. Help parties realistically evaluate alternatives to settlement.
8. Suggest that the parties take breaks when negotiations reach an impasse.
9. Encourage flexibility and creativity.
10. Shift the focus from past to future.
11. Shift the focus from one of blame to a creative exchange between the parties.
12. Hold caucuses with each disputant if there is deadlock or a problem.
13. Propose solutions that meet the fundamental interests of all parties.

Successful mediations result in a signed agreement or contract which is often called a memorandum of understanding and when signed becomes legally binding.

17.3.2.3 Conciliation

Conciliation is similar to mediation and both terms can be used interchangeably unless defined by contract. Even though the two terms have a number of similarities, there are also some differences between conciliation and mediation. In both cases, a neutral third party seeks to help the parties to settle their disputes amicably, the major difference between conciliation and mediation is determined by the amount of power the third party has. The conciliator, not the parties, often develops and proposes the terms of settlement between the parties, meaning a Conciliator usually offers his opinion on the relative strengths of the case, whereas in mediation, the Mediator facilitates a discussion between the parties and does not offer an opinion on the strength of each side's argument.

In Canada, the conciliation process is more commonly used to resolve disputes or conflicts between Employers and unions and between families.

17.4 Arbitration

If the parties cannot resolve their disputes through DAB or amicable settlement methods, arbitration is the next step in the construction dispute resolution process. Arbitration may be commenced any time prior to completion of the works or after such completion. As per FIDIC [2] Sub-clause 20.6, arbitration will be under the International Chamber of Commerce (ICC) arbitration rules if no alternative system of arbitration is stated.

Unlike the less formal process of mediation, the Arbitrator does not discuss the case individually with either of the parties prior to the hearing. The Arbitrator conducts a hearing, like in court, and issues a decision, known as an "award" that binds the parties. During the arbitration hearing, both parties have the right to present evidence and to introduce witnesses for substantiation in support of their positions. The Arbitrator has full power to review or revise any decision made earlier, and the Engineer can be called as a witness on matters related to the dispute.

The award must be made in writing and within the terms of reference; otherwise it will be invalid and therefore unenforceable. In general, arbitration proceedings often take many months and are therefore expensive.

Under most arbitration rules, the number of arbitrators could be a single Arbitrator or three arbitrators; however, FIDIC [2] conditions suggest appointing three arbitrators. A sole Arbitrator could be an expert in the domain of the dispute, or an expert in the law, or an expert in both, whereas a three-member panel usually preferred in large or complex cases could consist of:

(a) Three experts
(b) Two experts with the consent of parties and a lawyer as presiding Arbitrator
(c) Two lawyers with the consent of parties and an expert as presiding Arbitrator
(d) Three lawyers

It is recommended to select a presiding Arbitrator who is an expert in the law. The other two persons shall be experts in the domain of the dispute.

In Canada, arbitration is regulated by statute. Every province and territory has its own separate arbitration legislation. Each province and territory, with the exception of Quebec, has two arbitration statutes: (1) commercial and domestic arbitration (2) international arbitration. At the federal level, commercial arbitration is governed by the "Commercial Arbitration Act" (CAA). For more details, the "Dispute Resolution Reference Guide" published by Department of Justice Canada can be referenced, (which can be found on DOJ web site). The guide also provides sample agreement forms for parties requesting arbitration or mediation [6]. Some of the other organizations who have developed the arbitration rules are American Arbitration Association (AAA), ICC (International Chamber of Commerce) International Court of Arbitrators, ICC Canada Arbitration, ADR Institute of Canada, and Chartered Institute of Arbitrators.

17.5 Litigation

Litigation is usually the final procedure of a settlement and is utilized only when a construction claim cannot be resolved by a DAB or methods of amicable settlement or arbitration. Basically, a contract is a legally binding agreement; hence any disagreement or dispute can be referred to the courts of the country. Lawsuits are normally expensive and time consuming. Delays in settlement usually lead to heavy costs on daily basis. Claims litigation is very complex, and experience is essential. Good experienced lawyers and experts, however, also are very expensive.

Any party considering to pursue a delay or other claim should carefully evaluate the costs relative to any possible recovery. In addition, legal action can expose the parties to undesirable publicity. After a decision is rendered, the case is still subject to appeal or nonpayment that can result in additional time and cost.

Moreover, arbitration is conducted by technical people or legal people with experience in construction law and is generally conducted in the language of the project, whereas the courts may require that all the project paper work is translated into the language of the country.

References and Further Reading

1. CCDC-2. (2008). Canadian construction document committee – Stipulated price contract.
2. FIDIC. (1999). *Conditions of contract for construction*, Red Book: Author FIDIC (International Federation of Consulting Engineers).
3. CCDC 40. (2005). *Rules for mediation and arbitration of construction disputes*. Published by Canadian Construction Documents Committee.
4. Davies, J. G. (2011). Alternatives to the alternatives – A review of ADR procedures currently available to the Construction Industry in Canada- ADR Chambers Canada.
5. Totterdill, B. W. (2001). FIDIC (International Federation of Consulting Engineers) User's Guide; Thomas Telford Ltd.
6. Department of Justice Canada. (June 1995, updated March 2015). Dispute Resolution Reference Guide – Mediation.
7. Hollands, D. E. (1992). Amicable settlement of construction disputes – A report of FIDIC's Alternate Dispute Resolution Task Committee.

Chapter 18
Construction Contract Administration

18.1 Introduction

Many projects allocate more time to procurement than to administering the contract. This often leads to problems in Contractor performance, cost overruns, and delays in receiving goods and services. The real benefits of undertaking the procurement process can be achieved after the contract is awarded, so more attention needs to be placed on ensuring that the contract is implemented as intended and agreed.

In general, the contract administration role begins when design begins and ends when the project has been completed and accepted or it has been terminated and payment has been made after resolving the dispute. The early involvement of the Contract Administrator is valuable because there are many decisions that must be made during the preconstruction phase of the project. These decisions include choices that affect cost, procedures, and constructability. However, the main role of the Contract Administrator starts post award dealing primarily with the management of risks, monitoring performance to ensure that the objectives of the contract are met on time and within budget, and also detecting anything that might go wrong including suspension or termination of the contract.

Contract administration involves making decisions and the timely flow of information to enable completion of the project as required by the Contract Documents including review and observation of the construction project. This is important to all parties not only to determine that the work is proceeding in conformity with the Contract Documents but also because it ensures that their contractual obligations are fulfilled and legal rights protected.

Improving contract administration practices will help to achieve excellence in Contractor performance so that the buyer receives goods and services on time and within budget. Good contract administration assures that the end users are satisfied with the product or service being obtained under the contract.

A. Surahyo, *Understanding Construction Contracts*,
DOI 10.1007/978-3-319-66685-3_18

18.2 Contract Administrator

The Contract Administrator is the individual responsible for administering the construction contract (s). Contract administration involves numerous tasks occurring before and after contract execution. All work must be administered in accordance with the contract specifications, terms and conditions, provincial and federal laws and regulations, and department policy.

The Contract Administrator may be the Engineer or Architect, lead Consultant, the cost Consultant, or an Owner or Consultant's representative. This chapter will address the role of the Contract Administrator who acts as a representative of the Consultant.

Any construction project has many phases, and in most of the phases, the Contract Administrator's involvement is essential and valuable. Good contract administration is required to manage design specifications, contractual agreements, geotechnical investigations, product quality assurance, competitive tendering, evaluation, cost control, changes, final accounts, claims, and disputes. The Contract Administrator should possess thorough knowledge and understanding of the standard specifications, building codes and standards, project plans, construction techniques and methods, ability to understand and interpret the Contract Documents, and ability to communicate, negotiate, and resolve disputes and understand the administrative procedures established in the Contract Documents.

In general, the Contract Administrator's role includes:

1. Reviewing bid documents and processing bids
2. Receiving and evaluating and act on change proposals
3. Negotiating cost and schedule impacts related to change orders and other contract modifications
4. Reviewing and certifying payments under the Progress Payments clause
5. Monitoring progress for general conformity to the Contract Documents and prepare field reports of site observations
6. Issuing site or supplementary instructions when required
7. Seeking instructions from the Owner in relation to the contract
8. Processing shop drawings and request for information (RFI), coordinating receipt, tracking, and responses
9. Reviewing, evaluating, and responding to Contractor claims
10. Chairing construction progress meetings and preparing meeting minutes
11. Preparing and issuing construction progress reports
12. Preparing change order, contingency, RFI, and such other logs
13. Ensuring contractor compliance with quality assurance requirements
14. Monitoring commissioning and testing procedures
15. Processing disputes under the disputes clause
16. Ensuring project closing is handled properly and smoothly
17. Coordinating and communicating regularly with all stakeholders (contractors, owners, design team, and authorities having jurisdiction) to help achieve success for each project

These roles are further described under different project phases below.

18.3 Project Phases

Broadly, the project phases can be divided into three parts: preconstruction phase, construction phase, and post-construction phase.

18.3.1 Preconstruction Phase

It will be more convenient to understand the role of Contract Administrator throughout the preconstruction phase based on various stages:

- *Document stage:* Once the Consultant design team completes the project design based on the Owner's requirements and budget, the Consultant starts preparing the bid package which includes the contracting requirements, drawings, and specifications (Contract Documents). Prior to issuing Contract Documents for bidding, the Contract Administrator can add value by reviewing these documents. Many potential conflicts or deficiencies can be avoided by allowing the Contract Administrator, to review the Contract Documents as a final peer review. This will give the Contract Administrator opportunity to become familiar with the overall project and to find and resolve conflicts and deficiencies before they become issues during construction. This initial effort can help speed up construction and reduce construction costs.
- *Bidding stage*: After satisfactory review of Contract Documents by the Contract Administrator, the documents are issued for bidding. During the bidding stage, the Contract Administrator assists with processing Contractor inquiries, issuing of addenda and conducts the pre-bid conference to review the scope of work, contracting requirements, and project site conditions.

The Architect or Consultant is advisory to the Owner who will make the final decision to award a contract based on the bidding requirements. The Contract Administrator in conjunction with the Design Team will evaluate the bid submission, review the requirements listed in the instructions to Bidders, and verify proper bid forms are completed included with bid pricing. Ensure list of subcontractors and bidding Contractor's information is provided. Tabulating the bids and posting the results is also part of the Contract Administrator's responsibilities. After bid evaluation the Contract Administrator issues a recommendation to the Owner to award the contract to the successful bidder based on parameters established in the Contract Documents.

18.3.2 Construction Phase

The construction phase starts when the Owner issues a "notice to proceed" to the Contractor. This phase begins when the Contractor has entered into a formal agreement with the Owner. Construction is a team effort that includes contractors, subcontractors, testing agencies, designers, consultants, Owner's own forces, suppliers, and others, all working toward the common goal of completing the project ready for its intended use. The basic responsibilities of those involved in the construction process are stipulated in the Contract Documents. The construction phase involves many stages and is discussed as follows:

1. *Administrative Submittals:* Before commencement of construction activities, certain submittals are required to be provided by the Contractor. These establish a base line for site review efforts, cost analysis, and processes and procedures during construction. Administrative submittals may include the following project information or documents:

   - *Submittal schedule:* Frequently, the Contract Documents include a provision that obliges the Contractor to submit a schedule of submittals. Submittals are important communication tools for coordination and timely execution of projects. Submittals usually consist of shop drawings, samples, concrete and asphalt mix designs, construction schedules, quality control and safety plans, traffic control plans, and such other required documents as specified under the contract. Submittals are used to review the progress and material being used ensuring that everything is in agreement with the contract requirements.

     The Contractor prepares a submittal schedule reviewing each section of the Contract Documents and considers lead times needed for submittal preparation and reviews. The Contractor should add additional review time into the schedule for subcontractors and suppliers. The Contract Administrator's responsibility is to review time limits and encourage the Contractor to get submittals to the Consultant with sufficient lead time to avoid construction and delivery delays. The Contractor is also encouraged to provide the Contract Administrator with a master submittal list with anticipated time lines for advanced review. The Contractor should not proceed with fabrication or delivery of materials until related submittals are approved.
   - *WSIB certificates, bonds, andinsurance policies:* Before starting any construction on site, the Contractor has to submit required WSIB certificate, bonds, and insurance policies. The Contract Administrator has to receive these documents from the Contractor, review, and forward to the Owner for his review and record.
   - *Subcontractor lists:* If the Contractor has not submitted the subcontractors list for the project at bidding time, then the Contract Administrator must insist upon its submittal prior proceeding with the construction. If the Owner or the Consultant has any objections to any of the proposed subcontractors, the Contractor is notified in writing by the Contract Administrator. An alternative subcontractor must then be proposed, at no additional cost to the Owner. The Contract Administrator shall review and request confirmation of the subcontractor qualifications or certifications prior to acceptance of the final subcontractors list.

- *Schedule of values:* The Contractor is required to submit a schedule of values as the basis of the Contractor's applications for payment. The schedule of values breaks the work down into smaller cost-loaded portions of the work that become the basis for determining the percent of work complete. The schedule of values includes the total contract sum including allowance items and contract modifications. The schedule of values needs to be approved before an invoice is submitted by the Contractor. As construction proceeds, the schedule of values is updated to include amounts authorized by change order. On a regular basis the Contract Administrator shall review submitted payment applications against the approved schedule of values. The payment applications shall be verified for accuracy and for correct amounts reflecting actual percentage of work completed and materials stored at site or in store. The payment application is then forwarded to the Owner with a recommendation for payment as submitted or as adjusted.
- *Construction progress schedule:* Once the project is awarded, the Contractor prepares a proposed work schedule in accordance with the Contract Documents. The schedule is submitted to the Contract Administrator for review and acceptance. The main objective of the construction schedule is to document and communicate the Contractor's intended work plan and provide a baseline for monitoring progression of the work. Schedules should indicate the completion of the project within the allowable contract time. If the schedule does not reflect a reasonable or feasible plan to construct the project in the stipulated time period or the schedule is not prepared according to the specifications, the schedule will be returned to the Contractor for modifications.

The important aspects of a construction schedule is the contract performance periods, milestone events, submittal review/approval dates, product lead times, and activity durations. These activities in the construction schedule provide key information to the Consultant who will be scheduling time for field observations, submittal review, and project closeout activities. The regular project coordination meetings are an opportunity for the Contract Administrator to review construction progress and adherence to the schedule by the Contractor. The importance of maintaining the construction progress is to ensure delivery of the project on time.

The requirements of construction schedule are also emphasized in CCDC-2 [4] and FIDIC [8] forms of contract. CCDC-2 clause 3.5 requires that the Contractor shall submit a construction schedule to the Owner and Consultant before the first application for payment that indicates the timing of the major activities of the work with sufficient detail of the critical events, to demonstrate the work will be performed in conformity with the contract time. The Contractor is also required to update the schedule on a monthly basis or as specified in the Contract Documents informing the Consultant of any revisions required to the schedule as the result of extensions of the contract time.

Whereas FIDIC [8] Sub-clause 14.1 provides that within a set time of the letter of acceptance, the Contractor shall submit his program, for approval, in the form required by the Engineer. He shall also provide a written method statement as and when required by the Engineer. In continuity, the Sub-clause 14.2 specifies that if the Engineer considers that progress does not match the approved program, he may require the Contractor to produce a revised program showing how the works are to be completed on time.

2. *Preconstruction Meeting:* A preconstruction meeting is a valuable first step toward good communications in contract administration process. It is the first meeting with the project team and must be attended by all the major participants in the construction phase including Owner, Prime Consultant, Sub-consultants, Contractor, and major subcontractors. This meeting would highlight the base elements for the project and other project planning and scheduling activities. Preconstruction meeting introduces the members of the project team and the Owner and provides an opportunity to discuss the role of each team member.

This is a very important meeting and becomes critical to the success of this phase, by establishing the construction phase procedures and by identifying milestones, areas of special requirements, lines of communications and key contacts, submittal processes (shop drawings/samples), changes to the Contract Documents, payment procedures, and goals and objectives of the project. A successful preconstruction meeting will result in better communication and a more successful construction phase. If there are any issues to be raised regarding execution of the Construction Documents or the use of the site, it is during the preconstruction meeting that these issues are reviewed, and any restrictions are established. The Contractor is encouraged to disclose his initial mobilization efforts and any special requirements for set-up and planning construction.

An agenda should be prepared for all preconstruction meetings and sent to all invited parties with the notice of the meeting. The project size and complexity will determine the information to be covered in a preconstruction meeting. Some of the typical items that should be discussed at the preconstruction meeting include the following:

(a) Compliance with the Workplace Health and Safety Act and Regulations and any other applicable Acts
(b) Jobsite communication and authority
(c) Permits and licenses
(d) Insurance and bonding
(e) Quality control and assurance
(f) Work schedule and working hours of the Contractor
(g) Traffic accommodation
(h) Environmental requirements and issues
(i) Site office set-up
(j) Survey requirements
(k) Submittals procedures (shop drawings and samples)
(l) Progress payments
(m) Progress and other coordination meetings

(n) Any special requirements of the project
(o) Substantial completion dates, construction completion inspection and requirements for project completion

Meeting minutes must be documented promptly in order to be successful.

3. *Mobilization:* Mobilization refers to the activities carried out after the Owner has appointed the Contractor but before the Contractor commences work on site. It is a preparatory stage during which the majority of activities are managed by the construction team. Mobilization shall include all activities and associated costs for transportation of Contractor's personnel, equipment, and operating supplies to the site, establishment of site offices, and other necessary general facilities for the Contractor's operations at the site. The Contractor must do a basic site survey, establish property lines, and layout the site improvements and building locations. All this must be done to begin the site preparation work for the start of construction. It is during this post-bid period and initial project start-up phase that a number of issues are considered and resolved. The current site conditions are defined and overall project intent is reviewed. Subsequent to this, the Contract Administrator will observe the site preparations by the Contractor to determine if the contract limits have been observed as noted in the Contract Documents.

Verification shall be made that the site temporary facilities and offices have been correctly located, that temporary utilities have been extended to the site, that the required access control fences are accurately installed, and that staging areas have been developed in accordance with the Contract Documents. The Contract Administrator has to ensure that the access to and from the site is congruent with the drawings. The Contractor has to ensure that site access, site construction activities, limitations of working hours, and noise and dust control are in conformance with local authorities having jurisdiction.

4. *Monitoring Construction Work*: Consultants have certain responsibilities for making field observations during the construction process, and such duties are spelled out in the Owner-Consultant agreement. Hence, being the Consultant's representative, the Contract Administrator's main purpose of monitoring work on site is to become generally familiar with the progress and quality of the work completed and to ensure that the Contractor abides with the terms and conditions of the construction contract and that it follows the plans and specifications in constructing the project. Field monitoring and reporting to evaluate compliance of the work with the Contract Documents are among the most important responsibilities of the Contract Administrator and may include some or all of the following:

   (a) Observe the progress and quality of work for its conformity with the contract drawings and standard specifications, as well as record the necessary data to establish payment quantities under the schedule of tender quantities and unit prices or to make an assessment of the value of the work completed in the case of a lump sum price contract.

(b) Maintain complete and accurate records of the activities and events relating to the project, including date-/time-stamped photos.
(c) Ensure that all necessary safety precautions and protections are maintained by the Contractor throughout for the duration of the work.
(d) Review applications for payment, certify, and provide recommendations to the Owner.
(e) Interpret drawings, specifications, and special provisions for the project
(f) Obtain necessary approvals, and document significant changes to the project and maintain change order log.
(g) Attend or chair progress meetings at the site or a designated location.
(h) Review and take appropriate action upon the Contractor's submitted documents.
(i) Review and respond to requests for information (RFI) and other related Contractor submittals in coordination with other disciplines involved in the project, including structural, mechanical, electrical, and civil engineers, landscape architects, and other special consultants including maintaining RFI log.

Requests for information (RFI) are the vehicle for the Contractor to obtain additional or clarifying information when the Contract Documents are unclear, incomplete, and contradictory, or conditions vary from those depicted in the documents. It is important that the Contractor completely and clearly describe the problem, with a proposed solution when possible. It is also helpful to note the necessary time for response. The RFI becomes a document for the record to establish the issue to be resolved:

(j) Reject work or material which does not conform to the Contract Documents. The Contract Administrator should develop a process for advising the Contractor of nonconforming work which includes a time limit for implementing corrective measures. Defective work that has been rejected by the Contract Administrator shall be removed promptly from the work by the Contractor and replaced or re-executed according to the Contract Documents at no additional cost to the Owner.
(k) Resolve disputes which may arise in relation to the contract.
(l) Prepare and participate in dispute resolution or litigation regarding the project.

The Contract Administrator's site observation should not be confused with the supervision exercised by the contractor's staff. The Contract Administrator's principal duties on the job site are to observe, evaluate, and report, whereas the Contractor is responsible for controlling and directing the work and to arrange for observations by the Contract Administrator during key milestone periods. The frequency of observations may vary depending on the scope, scale, and phase of the project. Often, the general and supplemental conditions of the Construction Contract contain specific Consultant responsibilities during construction outlined in the Owner-Consultant agreement. The essential need for site visits is to document project status and any observed deficiencies.

The Consultant keeps the Owner informed of the progress and any deficiencies in the work. It is beneficial for the Contract Administrator to schedule more frequent site visits during critical milestone activities. These milestones are easily understood by reviewing the progress schedule and maintaining regular communication with the Contractor. It's essential to understand the limits of responsibilities of the Contract Administrator with respect to site visits. Consultant agreements and the general conditions state that the Consultant is not responsible for (1) construction means and methods, (2) project site safety, or (3) Contractor performance. The Contractor is contractually required to meet all these requirements, especially job site safety. The Contractor shall provide for safe access to the job site and the environment within the project limits whenever the Architect or Consultant visits the site.

5. *Record Keeping:* One of the major responsibilities of the Contract Administrator is to keep and maintain record of construction activities. Documentation is very important in a construction project because it provides a "memory" of the project. Clear record keeping plays an important role in communications and conflict avoidance. Documentation is the foundation on which all proposals, disputes, or claims are built. Without documentation, there is essentially no contemporaneous evidence, and, therefore, it is difficult to present a persuasive case. If all events and decisions are clearly recorded, it is easier to talk matters through when issues arise. Proper administration of the contract requires that all communications between the parties be in writing and preserved. Clear communication is vital to the success of a project, and written documentation of the communication process forms the basis for analysis and resolution of disputes whenever they arise. Typically, most of the following records are the standard construction project documents that Contract Administrator needs to maintain on a construction project:

    1. Original contract tender documents
    2. Issued for construction (IFC) drawings, revised drawings and As-Built drawings
    3. Site instructions issued to the Contractor
    4. Contemplated change notices issued by the Owner, change estimates, change requests and approved change orders including all relevant backup, computations, and change directives
    5. Change order logs and budget logs
    6. Shop drawings, shop drawing transmittals, and transmittals log
    7. Daily time sheets and diaries, covering weather conditions, time records, daily equipment and manpower used, general progress of the work, and an account of any extreme difficulties encountered by the Contractor accompanied with photographs
    8. Preconstruction photographs of existing facilities, site features, existing trees, and adjacent properties and should include the date taken and location and a description of the subject

9. Request for information record and log, dewatering logs, concrete pour log, and testing logs
10. Baseline construction schedule and updated schedule
11. Progress reports: weekly, monthly, or quarterly
12. Designer's inspection reports, testing agencies inspection reports, government inspector's inspection reports, and accident reports
13. Environmental issues, operational issues, and any such issues that may affect the construction completion schedule
14. Monthly payments issued to the Contractor
15. Preconstruction Meeting Minutes, Progress Meeting Minutes, and Coordination Meeting Minutes
16. Interoffice correspondence, Contractor correspondence, and Owner correspondence
17. Notice of substantial completion and notice of final completion including record of deficiency list
18. Notice of any claims and liens

18.3.3 Post-construction Phase and Project Closeout

As described in Chap. 11, the project closeout procedure passes through two important dates: date of substantial performance of the work and "total performance/Completion" date. On completion of the first stage of substantial performance of the work, the Owner can use the facility as it is considered ready for use as intended instead of waiting for total completion of the project which saves time and money.

The Contract Administrator's role becomes especially important during the construction closeout stage. The periodic field observations help to determine when the project is substantially complete to allow the Owner to use the site. It also identifies incomplete or nonconforming items, project completion dates, and when the Contractor is entitled to final payment. Substantial Completion is the agreed upon date when the project, or any portion thereof, is sufficiently complete that the Owner may use the project, or portion thereof, for its intended purpose. Final/total completion occurs when the Contractor has satisfied all the remaining items on the final deficiency list and the "Occupancy Permit" has been issued by the Authority having the jurisdiction including with final payment made by the Owner.

Primarily, the post-construction phase involves following stages:

Commissioning Commissioning is the process by which the completed facility is evaluated to determine that systems or components perform to the expectations of the Owner's requirements and the Contract Documents. Commissioning begins with the facility design concept but activated at the end of construction or at the point of substantial performance to successfully implement the start-up and operation of systems.

Commissioning is an additional service often provided by an independent third party – a commissioning agent. Large or complex projects may require the participation of a commissioning agent to manage and verify the design performance of all the components and systems of the facility's operation or as a minimum, those identified by the Consultant and Owner and specified to be critical or necessary. In some smaller and simpler projects, typical start-up demonstration and review by the Prime Consultant and Sub-consultant is usually sufficient.

Commissioning includes a range of activities undertaken to transform the design of a facility into a fully integrated and operating system. It is a process of quality assurance which [5]:

1. Begins with the definition of the "design intent" and ends with the delivery of a project
2. Confirms the Contractor's implementation of the Consultant's design as defined in the Contract Documents
3. Confirms the ability of the Consultant's design to satisfy the Owner's defined requirements
4. Addresses any shortcomings

Commissioning involves testing and recording of operational elements, verification of operations and maintenance requirements, and confirmation of planning required for future expansions or modifications. The Contract Administrator may be involved as the communications link and the records Manager transmitting final reports, deficiency lists, corrections required, and testing information.

Final Deficiency List Review As part of establishing substantial performance, it is necessary to place a value on project deficiencies. When the Contractor is of the opinion that the requirements of substantial performance as defined in Chap. 11 have been met, the Contractor shall make arrangements for an inspection of the work to be undertaken at the earliest opportunity and shall create a deficiency list of items to be completed or corrected. The Contract Administrator then attaches additional items if missed by the Contractor. This creates a final list of items to be corrected.

Deficiencies need to be completed as soon as possible to avoid delays in the process. Subcontractors should be involved early in the deficiency process, preparing their own deficiency lists. This will make the final review much easier for all parties. Subcontractors must also take responsibility for the quality of their work. The Consultant and Contractor must work together to judge the value of deficiencies to establish the value of works early. A standardized form shall be used for recording the deficient items which will assure all areas have been reviewed and that all items have been recorded.

When the Contractor has corrected the deficient items per the deficiency list, and has determined that the requirements for substantial performance of the contract have been met, the Contractor shall then make a written application to the Owner through the Contract Administrator for a certificate of substantial performance.

As discussed in Chap. 12, the application should include:

(i) The proposed date when the contract is considered to be substantially performed
(ii) The submission of all documentation required under the contract
(iii) A statement of cost values of work to be completed including correction of any deficient work
(iv) Any pending items which still need to be completed
(v) Any work which needs to be delayed to a later date if agreed between the parties
(vi) An invoice, claiming the basic holdback amount due, for release and payment following the issue of the certificate of substantial performance, including a Statutory Declaration and Workplace Safety and Insurance Board (WSIB) Certificate of Clearance,
(vii) Statutory Declaration Form [2] CCDC- 9A for subcontractor CCDC-9B is submitted

If the Contract Administrator determines that the contract has been substantially performed, the Contract Administrator shall certify the substantial performance of the contract by preparing and signing a certificate in Form 6 prescribed by the Construction Lien Act [3], specifying the date when the contract was substantially performed.

The Contractor then arranges to publish a copy of the certificate of substantial performance in a Construction Trade Newspaper and provides the Contract Administrator with proof of the date of publication, as required by the Ontario Construction Lien Act. The day following the date of publication shall be the date of commencement of the 45-day period prior to release of the basic statutory holdback monies.

The Contractor's and Contractor's subcontractors forces then continue to work toward final completion during the 45-day period mentioned. For further details, "OAA/OGCA (Ontario Association of Architects/Ontario General Contractors Association) Takeover Procedures [1]" can be referenced.

Final Completion When the Contractor is satisfied that the work is completed as required under contract and after making an inspection, the Contractor shall forward the inspection report and make a written request to the Contract Administrator for a review and assessment of the work. The Contractor's application as explained should include:

(i) The proposed date when the contract is considered to be finally completed
(ii) The submission of any balanced documents required under the contract
(iii) A statement of cost values for the work performed to the date of the completion
(iv) Any pending items which still need to be completed
(v) An invoice claiming for the finishing holdback, including a Statutory Declaration and Workplace Safety and Insurance Board (WSIB) Certificate of Clearance

If the Contract Administrator is satisfied that all deficiencies and uncompleted work have been corrected, he will forward the Contractor's invoice for final payment to the Owner for necessary payment. The Contract Administrator then prepares the statement of completion and the certificate for payment of the finishing holdback. This certificate shall be dated 1 day after the expiry of the 45 day period which commences on the day following the date the contract is considered as completed. After certificate for payment of finishing holdback is issued, the Contract Administrator shall advise the Owner to verify that no liens have been preserved as at the end of the 45-day period. After satisfaction that no lien has been preserved, the holdback for finishing work shall be paid 1 day after termination of the 45-day period.

Final completion is the end of all construction related activities, receiving of all submittals and extra stock and Owner final payment to the Contractor. The Contract Administrator is the facilitator of this process and verifies that all Operations and Maintenance Manuals have been received in the format required by the Contract Documents. The Contract Administrator also verifies all warrantees and guaranties have been received, all certificates have been issued, all commissioning reports are in and satisfactory, all deficiency list items are complete to Owner and Consultant's satisfaction, and final payments have been made.

The Owner will be relying on the closeout documents submitted by the Contractor such as "As-Built" drawings and manuals for as long as the structure is in use. Generally, closeout submittals include the following:

1. Record documents such as As-Built drawings and specifications, addenda, change orders, and site instructions.
2. Operation and Maintenance Manuals. Operation manuals provide information for the regular maintenance and cleaning instruction of materials, necessary instructions in order to operate systems and equipment. Maintenance manuals provide instructions regarding the cleaning methods and materials or regularly scheduled maintenance of equipment.
3. Spare materials, parts, and tools.
4. Product warranties.
5. Fire alarm certificate.
6. Balancing reports.
7. Testing certificates.
8. Occupancy permit.

18.4 Project Warranties

Warranties are used in construction projects to reduce financial risks associated with equipment and service deficiencies or failures related to construction activities. A warranty protects against costs related to construction deficiencies in materials or services for a specific period of time. Warranties can assist in holding contractors accountable for replacement or repair of construction deficiencies. For example, when a deficiency occurs during the warranty period for covered equipment such as an air conditioning unit or furnace, the unit is repaired or replaced at the Contractor's expense.

Guarantee and Warranty The difference between terms guarantee and warranty is always confusing. A simple difference is that a guarantee is an agreement between three or more parties and also known as collateral agreement. Like, a construction bond is a written agreement under which the surety (the guarantor) guarantees that the Contractor (the principal) will fulfill its obligations to the Owner.

On the other hand, a warranty is an agreement between two parties and is also known as bilateral agreement. Like an insurance policy, it is a contract between the insurer and the insured (known as the policy holder), under which the insurer promises, for a consideration, to assume financial responsibility for a specified loss or liability.

Almost every construction contract includes implied or express warranties. Warranties define and limit the responsibility of contractors for repairs of the construction project, both during and after completion of construction. Quality warranties identify the exact extent of what is covered by the warranty for the Contractor and time required for rectifying defects when reported. Whether the warranty is express or implied, each warranty can lead to liability for a party to the contract if the warranty is breached in some way.

18.4.1 Express Warranties

Express warranties are created and defined primarily by agreements between parties and the negotiated aspects of such agreements. Most construction warranties are express warranties and are contained in the Contract Documents. This warranty provides the Owner with the opportunity to provide written notice to the Contractor that certain work does not conform to the Contract Documents and, according to the conditions of the contract, the Contractor is required to correct the nonconforming work. A construction warranty usually protects the Owner against defects or failures within the warranty period also known as "maintenance period or defects liability period" and provides a remedy to the Owner for non-conformance with the contract discovered prior to or after substantial performance of the work. The warranty clause defines the responsibilities of all parties regarding the coverage of the warranty.

Most of the standard Contract Documents including CCDC-2 [4] recommends a 1-year warranty period. CCDC-2 [4] provides that the warranty period for the contract shall commence on the date of substantial performance of the work except extended warranties. Within this warranty period, the Contractor is contractually obliged to return to the construction site to repair defects which have appeared in the work. If the Owner finds any defect during this 1-year warranty period, he should immediately notify the Contractor and the Consultant. The Contractor should correct the defect promptly.

The major benefit to the Owner from the warranty period is that it provides an opportunity for rectifying defects which either do not need to be completed prior to substantial completion or which become apparent after substantial completion without the need for the parties to restore to dispute resolution.

18.4.2 Implied Warranties

Implied warranties arise by operation of law or imposed by law but are not expressly stated in the Contract Documents. They do not require any particular language or action by the parties to create them. It is an obligation that is implied by courts, arbitration panels, and/or dispute review boards created for equitable purposes.

Breach of an implied warranty is considered a Tort rather than a breach of contract. They are used to avoid injustice and ensure fairness between the contracting parties. The most common implied warranty claims relating to construction are:

1. Implied warranty of plans and specifications: Even though this provision is not spelled out in the contract, it is well established by the courts that the Owner's plans and specifications carry an implied warranty that they are accurate and suitable for the purposes of bidding and performing the contract.
2. Implied duty to provide access to the site: The Owner must have site ready for work, providing sufficient access to the site to allow contractors and subcontractors to accomplish their work and to allow for site investigation and inspection by contractors and subcontractors.
3. Implied duty not to prevent performance: It is the implied provision of most contracts that neither party to the contract should commit any act that will hinder or delay performance. A breach of this duty may result in delay damages and costs incurred because work was hindered, including extra manpower costs.

18.4.3 Extended Warranties

Extended warranties are usually provided by the warrantor to the benefit of the Owner. Any defect occurring during extended warranty period must be the responsibility of the warrantor or as specified in the Contract Documents. A number of Owners are now specifying 2 years to 20 years of extended warranties in their contracts.

Extended warranties are normally written by the manufacturer, for the manufacturer's benefit and therefore mostly contain the terms that limit the scope of coverage. Extended warranties usually exclude consequential and incidental damage to any building component other than the warranted product itself [7]. Even the most comprehensive extended warranties that cover material and workmanship generally provide that the warrantor will only repair faults that result from specific causes enumerated in the warranty. Sometimes manufacturer's warranty may restrict repair of the product failure only up to the cost of the original installation, not replacement cost. Warranties should be carefully read to determine exactly what is covered and what is excluded. The exclusions and limitations are very important.

Extended warranties in construction are typically of two types [6]:

- Product warranties and
- System warranties

18.4.3.1 Extended Product Warranties

Extended product warranties offered by product manufacturers usually cover the replacement of the product only, not the transportation, removal of the failed product, reinstallation of its replacement or any associated labor costs. For example, if you buy a submersible sewage pump, the manufacturer's warranty usually would state that the pumps are warranted to be free of material and workmanship defects for a 3-year period from the date of purchase. To obtain service for this sewage pump, you must return it, freight prepaid, to a service center authorized to repair these pumps. When requesting warranty service, you must present the proof of purchase documentation, which includes the date of purchase. During the term of this limited warranty, the authorized service center will repair any faulty workmanship and either repair or replace any defective part, at no charge to the original Owner.

However, the same warranty would state further that this warranty does not cover normal wear and tear or any malfunction, failure, or defect resulting from accident misuse, abuse, neglect, alteration, improper installation or maintenance, modification, and repair by other than a service center authorized to repair. Thus, the manufacturer's responsibility is usually limited to the supply of replacement product and will not include the removal of the existing product, installation of the new product, or any other incidental costs. CCDC contracts stipulate that the Contractor's responsibility with regard to extended product warranties is limited to obtaining the extended warranty documentation on the Owner's behalf from the manufacturer or supplier (warrantor). The extended product warranty documentation is to be issued by the manufacturer or supplier to the benefit of the Owner.

The other issue is that as time passes after the point of project completion, it becomes more and more difficult to discern between defects in the product and simple wear and tear.

18.4.3.2 Extended System Warranties

Extended system warranties cover replacement of the product and the entire installation, for example, roofing systems, glazing systems, and mechanical or electrical systems. These extended warranties are normally offered by manufacturers and cover components typically installed by a subcontractor approved by the manufacturer.

Extended system warranties usually contain restrictive provisions which significantly limit the warrantor's liability and the Owner's recourse in the event of a failure. For example, a manufacturer's inverted roof system warranty may cover the cost of repairing leaks which occur as a result of defects in their roof system components or defects in installation for a period of 5 years from the date of substantial performance. Should a leak develop in the system, the manufacturer's responsibility normally would not include the costs for removal and reinstallation of roof top landscaping materials or the costs associated with repair of interior finishes damaged as a result of the roof leak. The warranty could require the payment of a fee by the Owner and may state that any claims under the terms of the warranty be made within a very restrictive time period. As with extended product warranties, extended system warranties are obtained by the Contractor from the warrantor and issued by the warrantor to the benefit of the Owner.

Benefits

There are many benefits resulting from an equitable extended warranty that may make it worth the extra cost to the Owner. These benefits include the prequalification of the installer by the manufacturer, the manufacturer's involvement in the construction process, and the extended protection against failure. For the extended warranties to be effective, the warrantor must be financially secure to cover its potential liability.

Reference and Further Reading

1. Takeover Procedures – Ontario Association of Architects and Ontario General Contractors Association (OAA/OGCA) – Document # 100.
2. CCDC. (2001). *Canadian construction document committee- statutory declaration forms 9A and 9B.*
3. Ontario Construction Lien Act R.S.O, 1990, Chapter C30- Last Amended 2010,c.16,sched 2,s.2- Consolidation Period: From July 1, 2011 to e-Laws currency Date Nov 26, 2013.
4. CCDC-2. (2008). *Canadian construction document committee – Stipulated price contract.*
5. A Guide to Project Closeout Procedures – A joint Publication of Ontario Association of Architects and Ontario General Contractors Association.
6. Construction Warranties – CCDC Bulletin 17, November 1998.
7. Course Notes from Construction Contract Administration attended by Author. Conducted by Construction Specifications Canada (CSC) – March 2014.
8. FIDIC. (1999). *Conditions of contract for construction*, Red Book: Author FIDIC (International Federation of Consulting Engineers).

Index

A
Acceleration
constructive, 191
directed, 191
voluntary, 191
Acceptance, 2–4, 15, 25, 26, 38, 80, 87, 95, 96, 102, 103, 121, 128, 143, 145, 163, 166, 169, 170, 177, 228, 229
Actual cost (AC), 45, 46, 48–51, 65, 66, 94, 134, 151, 153, 198, 200
Addenda, 62, 76, 81, 83, 84, 227, 237

B
Bidding procedures
electronic, 86, 89
negotiated, xviii, 86, 88
open, xvii, 86, 87
selective, x, 55, 87, 88
Bidding requirements, 80–86, 227
Bid documents, xvii, 21, 36, 42, 61, 75, 76, 79–88, 90–92, 94, 121, 127, 226
Bid evaluation, xix, 15, 79, 80, 87, 88, 90–95, 227
Bid forms, 80, 82, 83, 85, 86, 96, 227
Bid modifications, 83, 89
Bid queries, 82, 84
Bid shopping, 12, 13, 15
Bill of quantities, 1, 18, 36, 46, 47, 62, 68, 74–76, 81, 94, 111, 148, 203
Bonds
bid, xix, 85, 90, 119–121
labour & material, xvi, 62, 63, 120, 123
lien, 120, 123, 124
maintenance, 120, 124
performance, xiv, 21, 62, 95, 116, 119–122, 124, 130
Build-operate-transfer (BOT), 39
Build-own-operate (BOO), 39
Business identification number (BIN), 29

C
Cash allowances, xvi, 148
Changes
cardinal, xiii, 192
constructive, x, 192
directed, xix, 192
Claim avoidance, 201–204
Closed-circuit television (CCTV) inspection, 176, 177
Commencement date, xvii, 127, 128, 217
Commissioning, 17, 18, 158, 226, 234, 235, 237
Completion date, x, 42, 54, 85, 127–129, 131, 132, 145, 147, 191, 194, 231, 234
Conditions of Contract
AIA, xix, 70, 71
CCDC, x, xiii, xvii, xix, 2, 17, 19, 20, 22, 23, 25, 46, 47, 114, 210, 211, 215
EJCDC, ix, 70
FIDIC, ix, xvii, xix, 17, 19, 20, 22, 24–27, 47, 55, 67, 103, 114, 115, 128–131, 143–146, 148, 149, 207–212, 215, 217, 218
general, 1, 12, 18, 19, 23, 63–65, 67, 70, 71
infrastructure, 39, 69
JCT, ix, 69, 70
NEC, xviii, 68, 69
supplementary, 64, 65

A. Surahyo, *Understanding Construction Contracts*,
DOI 10.1007/978-3-319-66685-3

Consideration, 2, 4, 43, 54, 62, 91, 95, 107, 108, 238
Construction claims
  claim records, 196–198
  dealing with claims, 8, 194–196
  pricing of claims, 198–201
Construction Lien Act, xviii, 7, 8, 10, 123, 124, 129, 146, 236
Construction safety, x, 26, 113, 179–188
Consultant, xix, 1, 3, 17–25, 36, 37, 41, 42, 47, 56, 57, 59, 61, 64, 67, 75, 76, 79, 82–84, 89, 95, 103–105, 107, 114–116, 118, 119, 127, 128, 130, 132, 133, 144–151, 158, 161–163, 165–167, 169, 180, 191, 193, 195–197, 202, 203, 206, 210, 211, 213, 215–217, 221, 226–233, 235, 237, 239
Contingency allowance, x, 148
Contract
  contract A, xv, 80, 121
  contract B, x, 80
  definition, 1, 18, 64, 143
  discharge, 24, 146, 147, 205–207
  Documents, ix, 1, 2, 11, 16, 18, 21–26, 36, 44, 47, 61, 62, 64, 68, 70, 71, 75, 76, 80, 81, 83, 87, 111, 121, 128, 131, 148, 150, 157, 162–169, 179, 193, 194, 201, 202, 214, 215, 225–232, 234, 235, 237–239
  suspension, x, 20, 25, 194, 205, 212, 213, 225
  termination, xvii, 15, 20, 25, 122, 127, 145, 194, 205–213
Contract Administrator, x, 51, 132, 144, 174–176, 191, 197, 215, 225–229, 231–237
Contract Documents, ix, 1, 2, 11, 16, 18, 21–26, 36, 44, 47, 61, 62, 64, 68, 70, 71, 75, 76, 80, 81, 83, 87, 111, 121, 128, 131, 148, 150, 157, 162–169, 179, 193, 194, 201, 202, 214, 215, 225–232, 234, 235, 237–239
Contract forms, 11, 62–64, 69, 80
Contractor, 1, 17, 29, 36, 45, 54, 61, 79, 97, 107, 127, 143, 161, 179, 189, 205, 215, 225
Contra proferentem, 16
Corporations, 4, 29–34, 137
Cost
  direct, 45, 48, 151, 152, 198–202
  fixed, 3, 49, 152
  impact, 54, 151, 152, 198–200
  indirect, 45, 94, 151, 152, 198–202
  variable, 152
Cost reimbursable contracts
  cost plus fixed fee, 48, 66
  cost plus percentage fee, 48, 66
  cost plus variable fee, 48, 49
  guaranteed maximum price, 48, 49, 66, 103
  time and material, 48, 51, 151, 197, 202
Critical path, xv, 131, 132, 136, 140, 141, 197

D

Deficiencies, 8, 12, 22, 57, 88, 104, 129, 146, 158, 163, 165, 177, 227, 232–237
Delays
  compensable, xviii, 190, 191
  concurrent, xix, 192
  excusable, x, 58, 190, 191, 194
  non-compensable, xviii, 190, 191
  non-excusable, xvi, 190, 191
Design-bid-build method, 36, 37, 55, 56, 70, 71
Design-build (DB) method, 36–38, 54, 55, 57
Design-build-finance-maintain (DBFM), 39
Design-build-maintain (DBM), 39
Differing site conditions, x, 58, 99, 100, 130, 192–195, 203
Dispute resolution
  amicable settlement, 220–224
  arbitration, x, 23, 215–217, 221, 223, 224
  conciliation, x, 222, 223
  dispute adjudication board, x, 23, 217–220, 223, 224
  litigation, x, 224
  mediation, x, 23, 215, 216, 221, 222
  negotiations, xiii, 23, 215, 216, 220, 221
Disruptions, xiv, 89, 119, 152, 194, 198, 200
Drawings, 1, 21, 22, 36, 44–47, 62, 72, 74–76, 81, 82, 94, 100, 103, 104, 111, 130, 132, 137, 138, 149, 158, 161, 163, 165, 167, 190, 193, 195, 197, 202, 226–228, 230–233, 237

E

Earned value technique (EVT), 152, 153, 155
Exclusion clause, 13–15, 109
Express terms, 11
Extension of time, xiii, 25, 127, 128, 130, 131, 150, 190–192, 194, 195, 212, 213, 229

F

Final payments, xiii, 145–148, 234, 237
Float
  free, 140

total, 140
Force majeure, xiii, 113, 114, 131, 194, 205, 207, 208, 210
Forecasting, 155
Form of Agreement, 45, 62, 71, 128
Forms of contracts, xv, xviii, 17, 23, 38, 55, 65, 67, 69, 103, 105–107, 110, 111, 114, 116, 127, 130, 132, 149, 150, 189, 195, 201, 212, 229

H
Holdback
basic holdback, 8–10, 146, 147, 236
finishing holdback, 8, 9, 147, 236, 237

I
Impact costs, 151, 152, 198–200
Implied terms, 12
Insurance
CAR, 117, 118
general liability, 111, 114–116, 118
indemnity, 104, 118, 119
policy, 21, 24, 26, 107, 108, 110, 111, 113–118, 130, 228, 238
property, 114–117

J
Joint venture (JV), 14, 29, 33, 34, 37, 67, 95, 103

L
Law
contract, 1, 2
environmental, 10
law of Tort, xiv, 2, 5–7
lump sum contract, 45, 46, 54, 58, 65, 144, 231
Private, 2
Public, 2
Leak testing
sewer, 47, 174–176, 184
watermain, 47, 172–174, 176, 184
Life cycle costing (LCC), 40, 156–159
Liquidated damages, 85, 129, 131, 191, 198, 207

M
MasterFormat, ix, 74
Mobilization, 230, 231
Mutual intent, 2, 4

N
Negligence torts, 6
Negotiated bidding, xix, 79, 88

O
Offer, ix, 1–3, 13, 15, 20, 32, 33, 37, 39, 54, 63, 79, 88, 91, 98, 117, 120, 122, 138, 157, 208, 223, 240
Owner, 1, 17, 29, 35, 45, 53, 61, 79, 97, 107, 128, 143, 165, 180, 189, 205, 215, 226

P
Partnership, 29–31, 33, 34, 36, 50
Preconstruction meeting, 230, 231
Privilege clause, 15, 80
Privity of contract, 5, 7
Professional indemnity insurance, 104, 118, 119
Progress payments, xviii, 19, 20, 143–145, 148, 150, 152, 153, 213, 226, 230
Progress schedule, 229, 233
Project delivery methods, 35, 70
Project phases
construction, xvii, 18, 44, 66, 157, 161, 227–234
post construction, 21, 227, 234–237
pre-construction, 21, 59, 66, 225, 227
Project scope, 53, 54, 140
Project warranties, 237–241
Public-private partnership (P3), 36, 38–40

Q
Quality management
quality assurance, 26, 161, 163, 165–177, 226, 230, 235
quality control, 24, 46, 158, 161, 163–177, 228, 230
quality planning, 163, 164
Quantum meruit, 11, 209

R
Risk management
risk analysis, 97–106, 158
risk identification, 59, 98, 99, 101, 103–105
risk monitoring, 98, 105

Risk management (*cont.*)
  risk planning, 98, 99
  risk response, 102–105

S

Safety, x, 19, 21, 22, 24, 26, 36, 57, 58, 62, 63, 91, 95, 101, 113, 147, 158, 161, 173, 175, 179, 189, 209, 210, 212, 228, 230, 232, 233
Schedule of rates, 47, 50, 62, 75, 203
Scheduling methods
  bar chart, 133–138, 197
  critical path method, xviii, 133, 137–141
  Microsoft Project, x, 133, 135–137
Single and multiple prime contract, 42–44
Sole proprietorship, 21, 29–31, 33
Specifications, ix, 1, 12, 21, 36–39, 42, 44–47, 62, 65, 67, 71–76, 79, 81, 82, 88, 94, 104, 110, 111, 118, 157, 161–163, 165–168, 172, 177, 180, 193, 194, 197, 201–203, 206, 226, 227, 229, 231, 232, 237, 239
Statutory declaration, 62, 147, 236
Subrogation, 110, 116
Substantial performance, 8, 9, 22, 121, 128, 129, 146, 147, 234–236, 238, 241

T

Termination, xviii, 15, 20, 25, 34, 113, 117, 120–122, 127, 145, 147, 158, 174, 191, 194, 205–214, 216, 218, 225, 237
Total performance, 129, 234
Types of contracts, xvi, 42, 43, 45–49, 57–59, 67, 68, 80, 103, 202

U

Umbrella liability, 111
UniFormat, xix, 74
Unit price contracts, 45, 47, 58, 65, 86, 144

V

Value engineering, 41, 156, 157
Vicarious liability, 6

W

Warrantees
  express, 238, 239
  extended, 238–241
  implied, 194, 238, 239
Workers safety insurance board (WSIB), 62, 63, 95, 228, 236

The manufacturer's authorised representative in the EU is Springer Nature Customer Service Centre GmbH, Europaplatz 3, 69115 Heidelberg, Germany. If you have any concerns regarding our products, please contact ProductSafety@springernature.com

Printed and bound by CPI Group (UK) Ltd, Croydon, CR0 4YY
30/09/2026
02225593-0003